PUBLIC HEALTH IN THE 21ST CENTURY

DIRECT-TO-CONSUMER GENETIC TESTS

CONSIDERATIONS AND QUESTIONABLE PRACTICES

PUBLIC HEALTH IN THE 21ST CENTURY

Additional books in this series can be found on Nova's website under the Series tab.

Additional E-books in this series can be found on Nova's website under the E-books tab.

GENETICS - RESEARCH AND ISSUES

Additional books in this series can be found on Nova's website under the Series tab.

Additional E-books in this series can be found on Nova's website under the E-books tab.

PUBLIC HEALTH IN THE 21ST CENTURY

DIRECT-TO-CONSUMER GENETIC TESTS

CONSIDERATIONS AND QUESTIONABLE PRACTICES

TREVOR HECHT
AND
AUSTIN F. MAZE
EDITORS

Nova Science Publishers, Inc.
New York

Copyright © 2012 by Nova Science Publishers, Inc.

All rights reserved. No part of this book may be reproduced, stored in a retrieval system or transmitted in any form or by any means: electronic, electrostatic, magnetic, tape, mechanical photocopying, recording or otherwise without the written permission of the Publisher.

For permission to use material from this book please contact us:
Telephone 631-231-7269; Fax 631-231-8175
Web Site: http://www.novapublishers.com

NOTICE TO THE READER

The Publisher has taken reasonable care in the preparation of this book, but makes no expressed or implied warranty of any kind and assumes no responsibility for any errors or omissions. No liability is assumed for incidental or consequential damages in connection with or arising out of information contained in this book. The Publisher shall not be liable for any special, consequential, or exemplary damages resulting, in whole or in part, from the readers' use of, or reliance upon, this material. Any parts of this book based on government reports are so indicated and copyright is claimed for those parts to the extent applicable to compilations of such works.

Independent verification should be sought for any data, advice or recommendations contained in this book. In addition, no responsibility is assumed by the publisher for any injury and/or damage to persons or property arising from any methods, products, instructions, ideas or otherwise contained in this publication.

This publication is designed to provide accurate and authoritative information with regard to the subject matter covered herein. It is sold with the clear understanding that the Publisher is not engaged in rendering legal or any other professional services. If legal or any other expert assistance is required, the services of a competent person should be sought. FROM A DECLARATION OF PARTICIPANTS JOINTLY ADOPTED BY A COMMITTEE OF THE AMERICAN BAR ASSOCIATION AND A COMMITTEE OF PUBLISHERS.

Additional color graphics may be available in the e-book version of this book.

Library of Congress Cataloging-in-Publication Data

ISBN 978-1-61942-175-2

Published by Nova Science Publishers, Inc. † New York

CONTENTS

Preface		vii
Chapter 1	Direct-to-Consumer Genetic Tests: Misleading Test Results Are further Complicated by Deceptive Marketing and Other Questionable Practices *United States Government Accountability Office*	1
Chapter 2	Direct-to-Consumer Genetic Testing and the Consequences to the Public *Jeffrey Shuren*	35
Chapter 3	Testimony of James P. Evans MD, Ph.D, Hearing of the House Energy and Commerce Committee's Subcommittee on Oversight and Investigations, July 20, 2010 *James P. Evans*	45
Chapter 4	Genetic Testing: Scientific Background for Policymakers *Amanda K. Sarata*	51
Chapter Sources		65
Index		67

PREFACE

In 2006, the Government Accountability Office (GAO) investigated companies selling direct-to-consumer (DTC) genetic tests and testified that these companies made medically unproven disease predictions. Although new companies have since been touted as being more reputable, experts remain concerned that the test results mislead consumers. This book examines the genetic tests currently on the market; the misleading test results themselves; the deceptive marketing techniques utilized; and other questionable practices

Chapter 1- In 2006, GAO investigated companies selling direct-to-consumer (DTC) genetic tests and testified that these companies made medically unproven disease predictions. Although new companies have since been touted as being more reputable—*Time* named one company's test 2008's "invention of the year"—experts remain concerned that the test results mislead consumers. GAO was asked to investigate DTC genetic tests currently on the market and the advertising methods used to sell these tests.

GAO purchased 10 tests each from four companies, for $299 to $999 per test. GAO then selected five donors and sent two DNA samples from each donor to each company: one using factual information about the donor and one using fictitious information, such as incorrect age and race or ethnicity. After comparing risk predictions that the donors received for 15 diseases, GAO made undercover calls to the companies seeking health advice. GAO did not conduct a scientific study but instead documented observations that could be made by any consumer. To assess whether the tests provided any medically useful information, GAO consulted with genetics experts. GAO also interviewed representatives from each company. To investigate advertising methods, GAO made undercover contact with 15 DTC companies, including the 4 tested, and asked about supplement sales, test reliability, and privacy policies. GAO again consulted with experts about the veracity of the claims.

Chapter 2- Dr. Jeff Shuren is the Director of the Center for Devices and Radiological Health (CDRH or the Center) at the Food and Drug Administration (FDA or the Agency). FDA's recent activities related to direct-to-consumer (DTC) genetic tests and the authors' future plans for the regulation of laboratory-developed tests.

Scientific advances resulting from the Human Genome Project completed in 2003 have expanded the authors' understanding of the genetic contribution to health and disease. These advances have also resulted in the development of new tests that can better identify individuals at risk for particular medical conditions and target medical treatments based on the likelihood that a patient will respond or experience an adverse event based on their individual genetic profile. FDA supports the promise and development of innovative genetic tests.

Chapter 3- Dr. James P. Evans is a physician and scientist who specializes in medical genetics.His research involves the use of emerging technologies to analyze the human genome for genes involved in cancer predisposition and the ways in which people use genetic information. He is the Editor-in-Chief of *Genetics in Medicine*, the official journal of the American College of Medical Genetics. But first and foremost he is a physician. He is a board certified internist who has a general medical practice. He is also board certified in Clinical Medical Genetics and in Molecular Genetic Diagnostics in which capacity he sees and tests patients who have, or are at risk of having, genetic disorders such as predisposition to cancer.

The breathtaking pace of discovery in the field of genetics is providing new opportunities for rapidly and inexpensively analyzing the human genome. One is now able to routinely query an individual's genome at over 1 million sites and the "$1,000 genome", in which access to one's entire genetic code will be feasible for many individuals, will soon be a reality.

Chapter 4- Congress has considered, at various points in time, numerous pieces of legislation that relate to genetic and genomic technology and testing. These include bills addressing genetic discrimination in health insurance and employment, personalized medicine, the patenting of genetic material, and the quality of clinical laboratory tests, including genetic tests. The focus on these issues signals the growing importance of the public policy issues surrounding the clinical and public health implications of new genetic technology. As genetic technologies proliferate and are increasingly used to guide clinical treatment, these public policy issues are likely to continue to garner considerable attention. Understanding the basic scientific concepts underlying genetics and genetic testing may help facilitate the development of more effective public policy in this area.

Most diseases have a genetic component. Some diseases such as Huntington's Disease are caused by a specific gene. Other diseases, such as heart disease and cancer, are caused by a complex combination of genetic and environmental factors. For this reason, the public health burden of genetic disease is substantial, as is its clinical significance. Experts note that society has recently entered a transition period in which specific genetic knowledge is becoming critical to the delivery of effective health care for everyone. Therefore, the value of and role for genetic testing in clinical medicine is likely to increase significantly in the future.

In: Direct-to-Consumer Genetic Tests
Editors: T. Hecht and A. F. Maze

ISBN: 978-1-61942-175-2
© 2012 Nova Science Publishers, Inc

Chapter 1

DIRECT-TO-CONSUMER GENETIC TESTS: MISLEADING TEST RESULTS ARE FURTHER COMPLICATED BY DECEPTIVE MARKETING AND OTHER QUESTIONABLE PRACTICES

United States Government Accountability Office

WHY GAO DID THIS STUDY

In 2006, GAO investigated companies selling direct-to-consumer (DTC) genetic tests and testified that these companies made medically unproven disease predictions. Although new companies have since been touted as being more reputable—*Time* named one company's test 2008's "invention of the year"—experts remain concerned that the test results mislead consumers. GAO was asked to investigate DTC genetic tests currently on the market and the advertising methods used to sell these tests.

GAO purchased 10 tests each from four companies, for $299 to $999 per test. GAO then selected five donors and sent two DNA samples from each donor to each company: one using factual information about the donor and one using fictitious information, such as incorrect age and race or ethnicity. After comparing risk predictions that the donors received for 15 diseases, GAO made undercover calls to the companies seeking health advice. GAO did not conduct a scientific study but instead documented observations that could be made by any consumer. To assess whether the tests provided any medically useful information, GAO consulted with genetics experts. GAO also interviewed representatives from each company. To investigate advertising methods, GAO made undercover contact with 15 DTC companies, including the 4 tested, and asked about supplement sales, test reliability, and privacy policies. GAO again consulted with experts about the veracity of the claims.

What GAO Found

GAO's fictitious consumers received test results that are misleading and of little or no practical use. For example, GAO's donors often received disease risk predictions that varied across the four companies, indicating that identical DNA samples yield contradictory results. As shown below, one donor was told that he was at below-average, average, and above-average risk for prostate cancer and hypertension.

GAO's donors also received DNA-based disease predictions that conflicted with their actual medical conditions—one donor who had a pacemaker implanted 13 years ago to treat an irregular heartbeat was told that he was at decreased risk for developing such a condition. Also, none of the companies could provide GAO's fictitious African American and Asian donors with complete test results, but did not explicitly disclose this limitation prior to purchase. Further, follow-up consultations offered by three of the companies failed to provide the expert advice that the companies promised. In post-test interviews with GAO, each of the companies claimed that its results were more accurate than the others'. Although the experts GAO spoke with believe that these tests show promise for the future, they agreed that consumers should not rely on any of the results at this time. As one expert said, "the fact that different companies, using the same samples, predict different directions of risk is telling and is important. It shows that we are nowhere near really being able to interpret [such tests]."

Gender	Age	Condition	Company 1	Company 2	Company 3	Company 4
Male	48	Prostate cancer	Average	Average	Below average	Above average
		Hypertension	Average	Below average	Above average	Not tested

Source: GAO.

Contradictory Risk Predictions for Prostate Cancer and Hypertension.

GAO also found 10 egregious examples of deceptive marketing, including claims made by four companies that a consumer's DNA could be used to create personalized supplement to cure diseases. Two of these companies further stated that their supplements could "repair damaged DNA" or cure disease, even though experts confirmed there is no scientific basis for such claims. One company representative even fraudulently used endorsements from high-profile athletes to convince GAO's fictitious consumer to purchase such supplements. Two other companies asserted that they could predict in which sports children would excel based on DNA analysis, claims that an expert characterized as "complete garbage." Further, two companies told GAO's fictitious consumer that she could secretly test her fiancé's DNA to "surprise" him with test results—though this practice is restricted in 33 states. Perhaps most disturbing, one company told a donor that an above average risk prediction for breast cancer meant she was "in the high risk of pretty much getting" the disease, a statement that experts found to be "horrifying" because it implies the test is diagnostic. To hear clips of undercover contacts, see http://www.gao.gov/products/GAO-10-847T. GAO has referred all the companies it investigated to the Food and Drug Administration and Federal Trade Commission for appropriate action.

Mr. Chairman and Members of the Subcommittee:

Thank you for the opportunity to discuss our follow-up investigation of genetic tests sold directly to consumers via the Internet. Using kits at home, consumers simply swab their cheeks or collect saliva and send these DNA samples back to a company for analysis and a report of the results. While the importance of genetics in individual medical care shows promise for the future, the usefulness of the tests these companies offer is much debated.

In 2006, we investigated four companies selling direct-to-consumer (DTC) genetic tests that purported to use DNA to deliver personalized nutrition and lifestyle guidance. We testified before the Senate Special Committee on Aging that these companies misled consumers by providing test results that were both medically unproven and so ambiguous as to be meaningless.[1] For example, one of the results we received vaguely indicated that our DNA donor was at "significant risk of developing the age related conditions associated with elevated levels of DNA damage." Another stated that our donor had "faulty methylation patterns" that may lead to "an above-average risk for developing cardiac aging, brain aging, and cancer." And though some of the companies claimed that they would provide lifestyle advice based on a consumer's DNA, we found that they simply provided generally accepted health guidance linked to background information submitted by our donors on test questionnaires. Further, two of the companies we tested recommended costly dietary supplements that were in reality nothing more than inexpensive multivitamins available at any drug store.

As a result of these findings, in 2006 the Centers for Disease Control and Prevention (CDC) in conjunction with the Food and Drug Administration (FDA) and the Federal Trade Commission (FTC) issued alerts warning consumers to be wary of claims made by these types of DTC genetic testing companies. In October 2008, FTC again warned consumers that "no standards govern the reliability or quality of at-home genetic tests. The FDA and Centers for Disease Control and Prevention recommend that genetic tests be done in a specialized laboratory and that a doctor or counselor with specialized training interpret the results."

Table 1. Donor and Profile Information

Donor	Profile	Gender	Age	Race or Ethnicity	Selected Medical History Information
1	Factual	Female	37	Caucasian	Colon cancer
	Fictitious	Female	68	African American	Hypertension and diabetes
2	Factual	Female	41	Caucasian	Breast cancer, diabetes, and heart disease
	Fictitious	Female	19	Asian	Heart arrhythmias
3	Factual	Male	48	Caucasian	Asthma, non-melanoma skin cancer, and heart disease
	Fictitious	Male	69	African American	Auto-immune disorders
4	Factual	Male	61	Caucasian	Colon cancer, heart disease, and a trial fibrilation
	Fictitious	Male	53	Caucasian	Prostate cancer and hypertension
5	Factual	Male	63	Caucasian	Type 2 diabetes, Alzheimer's disease, and obesity
	Fictitious	Male	29	Hispanic	Asthma and thyroid and colon cancer

Source: GAO

Note: We did not alter the gender on the donors' fictitious profiles because we believed that this difference would have been easily identified by these companies.

Despite these warnings, several new DTC genetic test companies have been touted as being more reputable and medically accurate than those we tested previously; in 2008, *Time* magazine named one new company's test the "invention of the year." More recently, another company's plan to sell tests at retail pharmacies has drawn significant attention from the media and scientists. However, given the scientific evidence currently available, many experts remain concerned that the medical predictions contained in the results mislead consumers. In this context, you requested that we proactively test DTC genetic products currently on the market and the advertising methods used to sell these products to consumers.

To investigate DTC genetic products currently on the market, we purchased tests, for $299 to $999, from a nonrepresentative selection of four of the dozens of genetic testing companies selling kits to consumers on the Internet.[2] Using online search terms likely to be used by actual consumers, we identified and selected these companies because they were frequently cited as being credible by the media and in scientific publications and because they all provided consumers with risk predictions, accessible through secure Web sites, for a range of diseases and conditions.[3] Although their tests are not identical, all four companies' Web sites contain a variation of the statement that their tests help consumers and their physicians detect disease risks early so that they can take preventive steps to reduce these risks. They also note that their tests are not intended to provide medical advice or to treat or diagnose disease. We purchased 10 tests from each company (40 tests in total) to compare risk predictions for a variety of serious illnesses and determine whether the companies were consistent in their predictions. We selected for comparison 15 common diseases and conditions that were tested by at least three of the four companies: Alzheimer's disease, atrial fibrillation (a type of irregular heart beat), breast cancer, celiac disease (a chronic digestive problem caused by an inability to process gluten), colon cancer, heart attack, hypertension, leukemia, multiple sclerosis, obesity, prostate cancer, restless leg syndrome, rheumatoid arthritis, type 1 diabetes, and type 2 diabetes.[4]

As shown in table 1, we then selected five DNA donors and created two profiles for each donor, one using factual information about the donor and one using fictitious information, including age, race or ethnicity, and medical history.

For each donor, we sent two DNA samples (saliva or a cheek swab) to each company—one sample using the factual profile and one using the fictitious—to determine whether altering the donors' backgrounds had any effect on the companies' DNA analysis. Three of the four companies asked for age and race or ethnicity prior to purchase; only one asked for medical history information. We also made undercover telephone calls to the companies seeking additional medical advice for both our factual and fictitious donors. We then documented our observations on the test results and advice we received. It is important to emphasize that we did not conduct a rigorous scientific study; our observations are those that could be made by any consumer. To assess whether we received any scientifically based or medically useful information, we consulted with external experts in the field of genetics and incorporated their comments as appropriate. Our primary consultant was Dr. James Evans, the Director of Adult Genetics Services at the University of North Carolina and the Editor-in-Chief of *Genetics in Medicine,* the official journal of the American College of Medical Genetics. After we completed our proactive testing, we visited each company and interviewed representatives who were willing to speak with us. We did not notify the companies prior to these visits and did not specifically disclose the results of our undercover testing or reveal the identities of our donors or the other companies that we tested.

To investigate the advertising methods used to sell DTC genetic products, we reviewed the Web sites of a nonrepresentative selection of 15 genetic testing companies, including the 4 from which we purchased tests. We identified the companies by again using online search terms likely to be used by actual consumers. Posing as fictitious consumers, we made contact with these companies, both by phone and in person, seeking additional information about genetic testing. During these contacts, we asked a series of questions about the reliability and usefulness of test results, privacy policies regarding consumers' genetic information, and the sale of supplements or other products. To assess the accuracy and reasonableness of the marketing claims, we again consulted with external experts in the field of genetics. We also purchased supplements sold by one of the companies.

Our findings are limited to the individual DTC genetic test companies we investigated and cannot be projected to any other companies. We performed our work from June 2009 to June 2010 in accordance with standards prescribed by the Council of Inspectors General for Integrity and Efficiency.

TEST RESULTS ARE MISLEADING AND OF LITTLE USE TO CONSUMERS

The test results we received are misleading and of little or no practical use to consumers. Comparing results for 15 diseases, we made the following observations: (1) each donor's factual profile received disease risk predictions that varied across all four companies, indicating that identical DNA can yield contradictory results depending solely on the company it was sent to for analysis; (2) these risk predictions often conflicted with the donors' factual illnesses and family medical histories; (3) none of the companies could provide the donors who submitted fictitious African American and Asian profiles with complete test results for their ethnicity but did not explicitly disclose this limitation prior to purchase; (4) one company provided donors with reports that showed conflicting predictions for the same DNA and profile, but did not explain how to interpret these different results; and (5) follow-up consultations offered by three of the companies provided only general information and not the expert advice the companies promised to provide. The experts we spoke with agreed that the companies' claims and test results are both ambiguous and misleading. Further, they felt that consumers who are concerned about their health should consult directly with their physicians instead of purchasing these kinds of DTC genetic tests. See appendix I for comprehensive information on the test results we received for each donor.

Different companies often provide different results for identical DNA: Each donor received risk predictions for the 15 diseases that varied from company to company, demonstrating that identical DNA samples produced contradictory results. Specifically, in reviewing the test results across all four companies for the donors' factual profiles, we found that Donor 1 had contradictory results for 11 diseases, Donor 2 for 9 diseases, Donor 3 for 12 diseases, Donor 4 for 10 diseases, and Donor 5 for 9 diseases. Specific examples of these contradictory predictions are listed below; note that some of the diseases we compared were only tested by three of the four companies. To facilitate comparison among companies, we chose to use the terms "below average," "average," and "above average" to describe the risk predictions we received; the exact language used by each of the companies is reprinted in appendix I.

- For Donor 1, Company 1 predicted an above-average risk of developing leukemia, while Company 2 predicted a below-average risk, and Company 3 reported that she had an average risk for developing the disease. In addition, Companies 2 and 4 told the donor that her risk for contracting breast cancer was above average, but Companies 1 and 3 found her only to be at average risk. See figure 1.
- Companies 1 and 2 claimed that Donor 2 had an above-average risk of developing type 1 diabetes, while Company 3 reported that she was at below-average risk for the disease. Further, Company 2 predicted she was at above-average risk for restless leg syndrome, Company 1 claimed she was at below-average risk for the condition, and Company 4 found that she was at average risk. See figure 2.
- Company 4 claimed that Donor 3's risk of developing prostate cancer was above-average, Company 3 found that he was at below-average risk, and Companies 1 and 2 found that he was at average risk. For hypertension, Company 3 found that he had an above-average risk of developing the condition, Company 2 found that he was at below-average risk, and Company 1 found he was at average risk. See figure 3.
- Donor 4 was told by Companies 1 and 4 that he was at above-average risk for celiac disease, but Company 2 reported that he was only at average risk. In addition, Companies 1 and 4 found that he was at below-average risk for multiple sclerosis, while Companies 2 and 3 found that he was at average risk. See figure 4.
- For Donor 5, Companies 2 and 3 reported an above-average risk for heart attacks, and Companies 1 and 4 identified only an average risk. Company 2 found him to be at below-average risk[5] for atrial fibrillation, while Companies 1, 3, and 4 predicted an average risk. See figure 5.

These contradictions can be attributed in part to the fact that the companies analyzed different genetic "markers" in assessing the donors' risk for disease. As described in a recent article published in the science journal *Nature*, researchers determine which markers occur more frequently in patients with a specific disease by conducting "genome-wide association studies, which survey hundreds of thousands or millions of markers across control and disease populations."[6] DTC companies use these publicly available studies to decide which markers to include in their analyses, but none of the companies we investigated used the exact same markers in its tests. For example, Company 1 looked at 5 risk markers for prostate cancer, while Company 4 looked at 18 risk markers.

In our post-test interviews, representatives from all four companies acknowledged that, in general, DTC genetic test companies test for different risk markers and that this could result in companies having different results for identical DNA. When we asked the representatives whether they thought that any DTC genetic test companies currently on the market were more accurate than others, all claimed that their own companies' tests were better than those offered by their competitors. For example, Company 1 said that it offers consumers more information than other companies because its results are based on both preliminary research reports as well as clinical data. Company 2 claimed that other companies do not test for as many markers as it does and that while none of the companies are "wrong," using more markers is "probably more accurate." Company 2 also stated that disparate test results from different companies are "caused, in part, due to a lack of guidance from the federal government, CDC in particular." Company 3 similarly claimed to test for more markers than other companies and stated that its test is "the best." Company 3 also said that there is a movement within the DTC genetic test industry to standardize test results, but that such standardization is a work in progress. Finally, Company 4 claimed that it uses stricter criteria to select risk markers than other companies. Company 4 also told us that it has been involved in a collaborative effort with other DTC genetic test companies to develop standard sets of markers, but stated that there are many unresolved differences in philosophy and approach.

Gender	Age	Condition	Company 1	Company 2	Company 3	Company 4
Female	37	Leukemia	Above average	Below average	Average	Not tested
		Breast cancer	Average	Above average	Average	Above average

Source: GAO.

Figure 1. Selected Contradictory Risk Predictions for Donor 1.

Gender	Age	Condition	Company 1	Company 2	Company 3	Company 4
Female	41	Type 1 diabetes	Above average	Above average	Below average	Not tested
		Restless leg syndrome	Below average	Above average	Not tested	Average

Source: GAO.

Figure 2. Selected Contradictory Risk Predictions for Donor 2.

Gender	Age	Condition	Company 1	Company 2	Company 3	Company 4
Male	48	Prostate cancer	Average	Average	Below average	Above average
		Hypertension	Average	Below average	Above average	Not tested

Source: GAO.

Figure 3. Selected Contradictory Risk Predictions for Donor 3.

Gender	Age	Condition	Company 1	Company 2	Company 3	Company 4
Male	61	Celiac disease	Above average	Average	Not tested	Above average
		Multiple sclerosis	Below average	Average	Average	Below average

Source: GAO.

Figure 4. Selected Contradictory Risk Predictions for Donor 4.

Gender	Age	Condition	Company 1	Company 2	Company 3	Company 4
Male	63	Heart attack	Average	Above average	Above average	Average
		Atrial fibrillation	Average	Below average	Average	Average

Source: GAO.

Figure 5. Selected Contradictory Risk Predictions for Donor 5.

When we asked genetics experts if any of the companies' markers and disease predictions were actually more accurate than the others, they told us that there are too many uncertainties and ambiguities in this type of testing to rely on any of the results. Unlike well-established genetic testing for diseases like cystic fibrosis, the experts feel that these tests are "promising for research, but the application is premature." In other words, "each company's results could be internally consistent, but not tell the full story.... [because] the science of risk prediction based on genetic markers is not fully worked out, and that the limitations inherent in this sort of risk prediction have not been adequately disclosed." As one expert further noted, "the fact that different companies, using the same samples, predict different... directions of risk is telling and is important. It shows that we are nowhere near really being able to interpret [such tests]." We also asked our experts if any of our donors should be concerned if the companies all agreed on a risk prediction; for example, all four companies told Donor 1 she was at increased risk for Alzheimer's disease. The experts told us this consensus means very little because there are so many demographic, environmental, and lifestyle factors that contribute to the occurrence of the types of diseases tested by the four companies.

Risk predictions sometimes conflict with diagnosed medical conditions or family history: Four of our five donors received test results that conflicted with their factual medical conditions and family histories.[7] When we asked the experts about these discrepancies, they told us that the results from these DTC tests are not conclusive because the tests are not diagnostic, as is noted on all of the companies Web sites. Because risks are probabilistic by definition, it is very likely that consumers will receive results from these companies that do not comport with their knowledge of their own medical histories. However, one expert noted that the discrepancies between actual health and the predications made by these companies also serve to illustrate the lack of robustness of such predictive tests. Moreover, experts fear that consumers may misinterpret the test results because they do not understand such distinctions. For example, a consumer with a strong family history of heart disease may be falsely reassured by below-average risk predictions related to heart attacks and consequently make poor health choices. In fact, one expert told us that "family history is still by far the most consistent risk factor for common chronic conditions. The presence of family history increases the risk of disease regardless of genetic variants and the current genetic variants do not explain the familial clustering of diseases." Another expert stated that "the most accurate way for these companies to predict disease risks would be for them to charge consumers $500 for DNA and family medical history information, throw out the DNA, and then make predictions based solely on the family history information." Examples we identified include the following:

- Donor 2 has a family history of heart disease yet all four companies predicted that she was at average risk for having a heart attack. Donor 2 also has a family history of type 1 diabetes, but Company 3 reported that she was at below-average risk for the disease.
- Donor 3 has a family history of heart disease, but Companies 1, 2, and 3 reported that he was at average risk for having a heart attack and Company 4 reported he was at below-average risk.
- Donor 4 had a pacemaker implanted 13 years ago to treat atrial fibrillation. However, Company 1 and 2 found that he was at below-average risk for developing atrial fibrillation,[8] and Companies 3 and 4 claimed that he was at average risk. Donor 4 is also a colon cancer survivor, but Company 2 reported that he was at average risk of developing the disease.

- Donor 5 has Type 2 diabetes, but Companies 1, 2, and 3 indicated that he had an average risk of developing the disease. Donor 5 is also overweight, but all four companies found him to be at average risk for obesity.

In our post-test interviews, representatives from all four companies reiterated that their tests are not diagnostic, but they all believe that their tests provide consumers and their doctors with useful information. Specifically, Company 1 stressed that its tests empower consumers to recognize their risk of developing a health-related condition and then take the information to a doctor for further discussion. Company 2 emphasized that its tests provide consumers with the "incentive" to be "aggressive" about their health, while Company 3 said its goal is to "empower individuals with information to help them make necessary lifestyle changes." Similarly, Company 4 stated that its risk predictions are a useful first step in that they offer "something for the consumer and their physician to consider in deciding whether or when to proceed with more invasive or costly tests." However, experts we spoke with cautioned that most doctors are not adequately prepared to use DTC genetic test information to treat patients. In addition, experts noted that there is currently no data or other evidence to suggest that consumers have taken steps to improve their health as a result of taking DTC genetic tests. As one expert noted, "even if such information is found to be an especially effective motivator of behavioral change, we're in trouble... because for everyone you find who is at increased disease risk, you'll find another who is at decreased risk. So if this information is actually powerful in motivating behavior then it will also motivate undesirable behaviors in those found to be at low risk."

Fictitious profiles did not receive complete test results: Many of these studies the companies use to make risk predictions apply only to those of European ancestry. Consequently, our fictitious Asian and African American donors did not always receive risk predictions that were applicable to their race or ethnicity, although the companies either did not disclose these limitations prior to purchase or placed them in lengthy consent forms. The experts we spoke to agreed that these limitations should be "clearly disclosed upfront" and suggested that our fictitious donors try to get their money back. Companies 2 and 3 did give us a refund, but Company 1 refused and company 4 never responded to our request. In our post-test interviews, company representatives acknowledged that race and ethnicity do affect disease risk predictions, but that most genetic research has only been done on persons of European ancestry and therefore such individuals receive more accurate results.

Representatives from Company 1 also said that the company can provide only current information and that one of its primary goals is to expand upon this research by collecting DNA from as many persons as possible. Further, Companies 2 and 4 stated that they believe they communicate this limitation to consumers on their Web sites or in their test result reports, though our observations do not support this claim. Examples of the discrepancies we identified include the following:

- Company 1 provided Donor 1's fictitious African American profile with test results based on her race for just 1 of the 15 diseases we compared: type 2 diabetes. For the remaining diseases, Company 1 provided a risk prediction but included a disclaimer, such as "this result applies to people of European ancestry. We cannot yet compute more precise odds" for those of African American descent. However, Company 1 did not explicitly disclose the fact that African Americans would receive incomplete results prior to purchase, even though it did ask consumers to specify their ethnicity as part of the purchase process. The company only vaguely refers to any testing limitations on the first page of its consent form, which states that "gene/disease associations are typically based on ethnicity and the associations may not have been studied in many world populations and may not apply in the same or similar ways across populations."
- Company 2 claimed on its Web site that it had "better coverage [of genes] associated with the most important diseases for all ethnicities" than its competitors. However, the company provided Donor 2's fictitious Asian profile with test results for just 6 of the 15 diseases we compared. The company did not explain these discrepancies and did not disclose the testing limitations prior to purchase, even though it requested that consumers specify their race or ethnicity as part of the purchase process. The only references to these limitations are made in the "frequently asked questions" section and on page six of an eight-page service agreement, where the company notes that "the genetic result reported may in some cases only be applicable to a certain group of people, e.g. based on gender, ethnicity, lifestyle, family history etc. that you may or may not belong to."
- Company 3 sent Donor 3's fictitious African American profile results for just 3 of the 15 conditions we compared. The company did not

- disclose this limitation prior to purchase even though it requested that consumers specify their race or ethnicity during the purchase process.
- For 10 of the 15 conditions we compared, Company 4 sent all of our donors results that applied only to individuals of European ancestry. However, for restless leg syndrome, the predictions were accompanied by the following statement: "most conditions have only been studied in people of European ancestry. But this condition is a little different." For atrial fibrillation, colon cancer, type 2 diabetes, and heart attack, the predictions were accompanied by the following statement "most conditions have only been studied in people of European ancestry, but this one also has been studied in other groups." The company provided no additional explanation as to how these differences applied to our donors. The only other reference to testing limitations is made on page five of a nine- page consent form, where the company notes that "most of the published studies in this area of genetic research have focused on people of Western European descent. We do not know if, or to what extent, these results apply to people of other backgrounds."

Company 1 provided conflicting predictions for the same DNA within the same test result report: Company 1 provided our donors with conflicting risk predictions for atrial fibrillation, celiac disease, and obesity. In reviewing the test results for just the factual profiles, we observed the following:

- Donor 1 received a "clinical report" predicting that she had an average risk for developing atrial fibrillation and a "research report" stating that she was at below-average risk for the disease.
- Donor 2 received a "clinical report" stating that she was at below-average risk of developing celiac disease and a "research report" claiming that she was at above-average risk.
- Donor 4 received one "research report" claiming that was at above-average risk for obesity and another "research report" stated that he was at average risk.

According to information in the test results, the company distinguishes between clinical and research reports by noting that predictions based on the clinical reports are for "conditions and traits for which there are genetic associations supported by multiple, large, peer-reviewed studies." In contrast, the research reports provide information "that has not yet gained enough scientific consensus to be included in our clinical reports." However, there is no additional information explaining how consumers should interpret the results. Because the company does not offer any follow-up consultations on test results, our fictitious donors could not request clarification. When we interviewed representatives from Company 1 about this issue after our testing, they simply reiterated the information contained in the results, describing research reports as being peer reviewed and "almost clinical" but noting that clinical reports are "four star" in that they are widely accepted according to scientific standards.

Follow-up consultations provide only general information: As part of the test results, all four companies provide generally accepted health information related to the diseases that were tested, including a description of symptoms, treatments, and methods of prevention. This information is not targeted to specific consumers; all of our donors' results contained the same descriptions of treatments and methods of prevention, regardless of the risk predictions they received. For example, all the companies note that stopping smoking and increasing exercise are ways to reduce the risk for heart attacks. Representatives for Company 4 also encouraged consumers to make dietary changes such as adopting a Mediterranean diet or eating curry to prevent Alzheimer's disease, claims that cannot be proven, according to our experts. To supplement this information, Companies 2, 3, and 4 offer follow-up consultations.[9] Only Company 4 has U.S. board-certified genetic counselors on staff for this purpose, but all three companies claimed on their Web sites that their representatives would help consumers understand the implications of their disease risk predictions. However, for the most part, these representatives provided our donors with little guidance beyond the information contained in the test reports; at times, it seemed as though they were simply reading information directly from these reports. When our donors asked for more information on alarming results that indicated that they were at increased risk for serious diseases like colon cancer and Alzheimer's disease, representatives for Companies 2 and 3 pointed out symptoms to be aware of, but acknowledged that there is very little the donors could do to mitigate these risks. Representatives for Companies 2 and 4 also conceded that the donors' own doctors would probably not know what to do with the test results, a fact that our experts repeatedly noted. Examples include the following:

- Company 2 offers follow up consultations with "experts" to help consumers "interpret their results." In our post-test interviews, the company further noted that it provides the option of speaking with genetic counselors or a medical geneticist, but that consumers rarely exercise this option. Because the company is located outside the country, we were unable to determine whether all of its counselors are board certified in the United States; however, one counselor told us that he was not certified. During one of our undercover follow-up calls, Donor 4 asked what to do about his test results in general and what lifestyle changes he should make as a result. The representative told Donor 4 that he could not tell him what to do because he was not a physician and that the donor should take his results to a physician if he wanted advice on making any changes. When Donor 4 expressed concern that his doctor may not know what to do with the test results, the expert told him "True, not all physicians are familiar with these tests, so if you were to take it into a physician's office, they may not be familiar with it." Furthermore, when discussing Donor 3's increased risk for colon cancer, one of Company 2's experts told our donor that while he should become familiar with the symptoms such as blood in the stool, there was not much else he could do because "colon cancer is quite silent."

- Company 3 states that "because of the complexity and inherent uncertainties in genetic information, we recommend that you discuss the results of your genetic test with a genetic professional... .Our on-staff Genetic Counselors are available any time to review your... results with you." In our post-test interviews, the company further claimed that its genetic counselors are certified by the American Board of Genetic Counseling and that the counselors review family history and provide consumers with additional information that is not in the test results. However, our donors spoke to the same person, who admitted that she was not a board-certified genetic counselor. She told us that she had completed her master's in genetic counseling and just had to take her test to become licensed. Donor 5 called Company 3 because he was extremely concerned about the company's prediction that he had genetic markers that are highly correlated with Alzheimer's disease. Instead of providing addition information, the counselor simply acknowledged that "there is no cure or prevention strategy with Alzheimer's."

- Company 4 notes that its "genetic counselors are healthcare professionals who are trained to help you understand what genetic information means for you and for your family." In our post-test interview, the company stressed that its counselors explain the results, discuss beneficial next steps, and ensure that consumers and their physicians understand the meaning and limitations of the tests. However, when Donor 2 asked what she could do about her test results, the counselor told her that she could take the results to a physician. When Donor 2 pressed the counselor about whether a doctor would know what to do, the counselor responded "With this stuff? Probably not, no, I think they're learning just like everyone else."

"PERSONALIZED" SUPPLEMENTS, BOGUS ENDORSEMENTS, AND SCIENTIFICALLY INVALID CLAIMS AMONG DECEPTIVE MARKETING PRACTICES

Posing as consumers seeking information about genetic testing on the Internet and through phone calls and face-to-face meetings, we found that 10 of the 15 companies we investigated engaged in some form of fraudulent, deceptive, or otherwise questionable marketing practices. For example, at least four companies claimed that a consumer's DNA could be used to create personalized supplements to cure diseases. One company's representative fraudulently used endorsements from high-profile athletes to try to convince our undercover investigators to purchase its supplements. He also told our fictitious consumers that they could earn commission checks and receive free supplements if they could convince their friends to purchase the products. More detailed information on our experiences with this company follows table 2. Another flagrant example of deceptive marketing involved several companies' claims that they could predict which sports children would excel in based on DNA analysis. We also found examples of highly misleading representations about the reliability of the tests and the ability of health care practitioners to use the results to help treat patients. In addition, two companies are placing consumers' privacy at risk by condoning the potentially illegal practice of testing DNA without prior consent. Selected audio clips from our undercover calls and meetings are available at http://www.gao.gov/products/GAO-10-847T. Table 2 contains a selection of representations made by these companies. Note that companies 1 through 4 are the same companies we proactively tested, as discussed earlier in this testimony.

Company 5: On its Web site, Company 5 claimed that it would use a consumer's DNA to "create a personalized formula for nutritional supplements and skin repair serum with 100% active ingredients individually selected to enhance or diminish the biological processes causing you to age." To investigate these claims, we posed as a fictitious consumer interested in purchasing the product and met in person with a company representative.

Table 2. Examples of Deceptive Marketing, Misinformation, and Questionable Practices

Source	Representation	Comments
Company 5	Representative claimed Michael Phelps used the companies' supplements. Representative also claimed that he would be meeting with Lance Armstrong because his doctors thought that test was "the most amazing thing they've ever seen."	Representatives for Michael Phelps and Lance Armstrong told us that they had never heard of this product and had no endorsements or dealings with the company.
Company 5	Company representative claimed that use of the company's supplements cured the arthritis in his knee and prevented him from getting high blood pressure and high cholesterol. He also suggested that our fictitious consumer could stop taking his cholesterol medication once he started taking the company's supplements.	"Absolute lies," said one expert about these claims. Experts also stated that the claims have no scientific basis and consumers could suffer serious health consequences if they follow this advice. Moreover, FDA and the National Institutes of Health have noted that no dietary supplement can treat, prevent, or cure any disease.
Company 6	Genetic counselor claimed that tests and related products could help "repair damaged DNA."	Experts told us there is no scientific basis for this claim.
Companies 7 and 8	Companies claim to use a consumer's DNA and or genotype to create a "custom blend of nutrients" and "diet and exercise guidelines."	During a conversation with one of our fictitious consumers, a company representative admitted that supplements are just "high-quality vitamins and minerals" and that diet and exercise guidelines are merely

Source	Representation	Comments
		based on a consumer's responses to a questionnaire. Experts told us that there is no scientific basis for suggesting that supplements, diet, or exercise can be customized to DNA.
Companies 9 and 10	Web sites claim to be able to predict athletic performance by analyzing DNA and also to be able to determine which sports children will excel in.	"In unqualified terms, [these claims] are complete garbage," according to one expert.
Companies 1, 2, 3, and 4	Web sites and company representatives told us that consumers should bring test results to their physicians to be used as a "tool" for treatment.	According to the Department of Health and Human Services' Secretary's Advisory Committee on Genetics, Health, and Society, "[practitioners] cannot keep up with the pace of genetic tests and are not adequately prepared to use test information to treat patients appropriately." Therefore, direct to consumer genetic tests may not provide any substantial utility to the consumer.
Companies 4 and 9	Although their Web sites state that tests are not intended to diagnose diseases, a representative for Company 4 claimed that its tests were "diagnostic" and a representative for Company 9 claimed that its tests were "prognostic" when asked about their reliability.	Experts described these statements as "horrifying" and "disconcerting," because they could mislead consumers into thinking that they have a disease or provide a false sense of assurance that they don't. In addition, experts told us that for the types of conditions being tested by these companies, multiple studies have confirmed that DNA testing adds little to an analysis of a person's weight, age, gender, and family history.

Table 2. (Continued)

Source	Representation	Comments
Company 4	"You'd be in the high risk of pretty much getting it," is how a representative responded when our fictitious consumer asked if results indicating she was at above average risk for breast cancer meant she's definitely getting the disease.	Experts also called this statement "disconcerting" and "horrifying" because it erroneously implies that the test can diagnose breast cancer and could needlessly alarm consumers.
Company 6	In response to general inquiries about genes and genetic testing, a representative stated that "genes are a symptom not a source of our biology."	An expert characterized this statement as "nonsensical."
Companies 3 and 4	Although company Web sites require consumers to explicitly consent to genetic testing before submitting a DNA sample, representatives from these companies told our fictitious consumer that she could secretly send in her fiancé's DNA and "surprise" him with the results.	One expert characterized the companies' willingness to conduct tests without prior consent as "dangerous" and "irresponsible." According to the Johns Hopkins Genetic and Public Policy Center, this "surreptitious" testing could lead to people "learning of health risks or family relationships that he or she would prefer remain unknown." Currently 33 states place some type of restrictions on surreptitious testing.[a]

Source: GAO.

[a] For purposes of our testimony, surreptitious testing refers to the collection, analysis, or disclosure of the results of DNA samples without the consent of the person tested. State laws restricting surreptitious testing vary. For example, some states prohibit surreptitious testing for health-related purposes while other states restrict such testing for other purposes, including the determination of parentage. In a few states, the laws restricting surreptitious testing only apply to insurance companies.

During our initial meeting, the representative not only fraudulently suggested that Michael Phelps and representatives for Lance Armstrong endorsed the product, he also implied that the company's supplements could cure high cholesterol and arthritis, claims that one of our experts characterized as "absolute lies." Moreover, the FDA and the National Institutes of Health have clearly stated that no dietary supplement can treat, prevent, or cure any disease. As part of the company's promotional materials we found that the company's DNA assessment cost $225 and that the customized supplements cost about $145 per month. However, if our fictitious consumer immediately purchased a 3-month supply of supplements, she would be able to get the DNA test for free. The representative also told her that she could become a company affiliate and earn commission checks and free products by recruiting new affiliates. She, along with another fictitious consumer, subsequently registered as company affiliates, and ultimately received commission checks totaling more than $250. In addition to sending us the test kits, the company sent us packages of starter supplements in a bag that was not labeled with an ingredient list.

In an attempt to compare the test results from Company 5 with the results we received from Companies 1 through 4, we again used the same five donors and replicated the same methodology: submitting DNA samples using one factual profile and one fictitious profile. However, when we received the results, we found that Company 5 did not provide a set of risk predictions for specific diseases, making it impossible for us to compare the results against those we received from the other four companies. Instead, the company sent our donors a list of gene variants tested, a description of bodily functions affected by those variants, and a determination of whether the donors needed additional "nutritional support" to maintain health. In comparing the results, we found that each donor appeared to have a unique assessment and that using the fictitious profile did not seem to affect the results. However, the results were so ambiguous and confusing that they did not provide meaningful information. For example:

- Donor 1 was told that she needed "maximum support" to maintain the "VDR gene" which accounts for "75% of the entire genetic influence on bone density" among healthy people. Maximum support means that the "protein molecule expressing a specific enzyme, hormone, cytokine or structural protein is functioning minimally" and maximum nutritional support is needed to keep the body functioning optimally.
- Donor 5 was told that he needed "added support" to maintain the "EPHX" gene, which "detoxifies" epoxides or "highly reactive foreign chemicals present in cigarette smoke, car exhaust, charcoal-grilled meat, smoke from wood burning, pesticides, and alcohol."

"Added support" means that the gene is functioning less than optimally and therefore needs added nutritional support.

According to one of the experts we spoke with, these claims are simply "nonsensical" and "while it is true that one can find alleles[10] of many of these genes that don't have the same activity as 'normal,' we have no idea of (a) whether that reduced activity has any real health implications and (b) what one would reasonably do about it if so."

Along with the test results, the company sent supplements that it claimed were "blended" based on our donors' DNA assessments. The supplements arrived in the same type of unlabeled bag as the starter supplements. This time, the ingredients were printed inside the test result booklet sent to each donor and included substances such as raspberry juice powder, green tea extract, and garlic powder. The recommended daily dose is 10 supplements per day. Based on a review of all the ingredient lists, our five donors appeared to get supplements with different combinations of substances. However, we did not test the supplements to verify their contents. Moreover, an expert we spoke with told us that there is no scientific basis for claiming that supplements can be customized to DNA.

In post-test interviews, Company 5 told us that this company differs from others in that it does not attempt to diagnose or calculate a predisposition to any disease. Instead, the company said that it focuses on the overall health and well-being of their clients by creating personalized nutritional supplements based on their client's specific DNA. When we asked about the ingredients in the supplements, the company told us that all supplements have a base formula of ingredients that their scientists have determined to be "beneficial for everyone." Additional nutrients are then added to the base formula based on deficiencies identified by the company's DNA test. When we asked about the endorsements, we were told that several celebrities and professional athletes use the company's products, but that many of these high-profile clients do not want to disclose this affiliation.

Corrective Action Briefings

We briefed FDA, the National Institutes of Health, and FTC on our findings on May 25, 2010; June 7, 2010; and June 17, 2010, respectively. In addition, we have referred all the companies we investigated to FDA and FTC for appropriate action.

Mr. Chairman, this concludes my statement. I would be pleased to answer any questions that you or other members of the committee may have at this time.

APPENDIX 1: TEST RESULTS BY DONOR

This appendix provides (1) a description of both the factual and fictitious profiles used by each donor and (2) tables documenting the risk predictions we received from all four companies for the 15 diseases we compared.

To the extent possible, we have used in the risk prediction language directly from the test results. However, Company 2 did not use terms like "average" or "below average" to describe risk. Instead the company used charts showing each consumer's risk level as compared to others with the consumer's gender and ethnicity or as compared to those of European ancestry. The results were color coded, with green to light green appearing to correspond to a below-average risk level, yellow corresponding to an average risk level, and orange and red corresponding to an above-average risk level. To facilitate comparison, we chose to use these corresponding terms to describe the results, as shown in the table. In addition, Company 1 used two different types of reports in its test results: clinical and research. According to the company, the clinical reports contain "information about conditions and traits for which there are genetic associations supported by multiple, large, peer-reviewed studies." Research reports contain "information from research that has not yet gained enough scientific consensus to be included in our clinical reports." Where applicable, we noted when a risk prediction was derived from a research report; all the other predictions were derived from the clinical reports.

Donor 1: Donor 1 is a 37-year old Caucasian female, who eats a balanced diet and exercises regularly. She has elevated cholesterol and arthritis in her back. In addition, she has a strong family history of colon cancer and a grandparent who was diagnosed with dementia. In Donor 1's fictitious profile, she is a 68-year old, African American female, who is overweight and rarely exercises. She has type 2 diabetes, hypertension, and asthma, but has no family history of colon cancer or dementia.

Table 3. Comparison of Test Results for Donor 1

Disease or condition	Profile	Company 1	Risk predictions Company 2	Company 3	Company 4
Alzheimer's disease	Factual	Not tested	Above average	Increased susceptibility	Above average
	Fictitious	Not tested	Not tested	Increased susceptibility	Above average
Atrial fibrillation	Factual	Typical and decreased (research)	Average	Average predisposition	About average
	Fictitious	Typical and decreased (research)	Not tested	Not tested	About average
Breast cancer	Factual	Typical	Above average	Average predisposition	Greater than most women's
	Fictitious	Typical	Not tested	Average predisposition	Greater than most women's
Celiac disease	Factual	Decreased and typical (research)	Average	Not tested	Below average
	Fictitious	Decreased and typical (research)	Not tested	Not tested	Below average
Colon cancer	Factual	Elevated	Above average	Increased susceptibility	Above average
	Fictitious	Elevated	Not tested	Not tested	Above average
Heart attack	Factual	Decreased	Average	Average predisposition	Average
	Fictitious	Decreased	Not tested	Not tested	Average
Hypertension	Factual	Elevated (research)	Average	Average predisposition	Not tested
	Fictitious	Elevated (research)	Not tested	Not tested	Not tested
Leukemia	Factual	Elevated (research)	Below average	Average predisposition	Not tested
	Fictitious	Elevated (research)	Not tested	Not tested	Not tested

Disease or condition	Profile	Risk predictions			
		Company 1	Company 2	Company 3	Company 4
Multiple sclerosis	Factual	Decreased	Average	Average predisposition	Below average
	Fictitious	Decreased	Not tested	Not tested	Below average
Obesity	Factual	Typical and typical (research)	Below average	Average predisposition;	Below average
	Fictitious	Typical and typical (research)	Not tested	Not tested	Below average
Prostate cancer	Factual	Not applicable	Not applicable	Not applicable	Not applicable
	Fictitious	Not applicable	Not applicable	Not applicable	Not applicable
Restless leg syndrome	Factual	Decreased	Below average	Not tested	Below average
	Fictitious	Decreased	Not tested	Not tested	Below average
Rheumatoid arthritis	Factual	Decreased	Below average	Average predisposition	Below average
	Fictitious	Decreased	Not tested	Not tested	Below average
Type 1 diabetes	Factual	Elevated	Above average	Do not show strong susceptibility	Not tested
	Fictitious	Elevated	Not tested	Not tested	Not tested
Type 2 diabetes	Factual	Typical	Average	Average predisposition	Below average
	Fictitious	Typical	Below average	Average predisposition	Below average

Source: GAO analysis of results from four companies.

Donor 2: Donor 2 is a 41-year-old Caucasian female. She is in good health; however she has a family history of breast cancer, type 1 diabetes, and heart disease. In Donor 2's fictitious profile, she is a 19-year-old Asian female who smokes, drinks and uses recreational drugs. She suffers from heart arrhythmias and an elevated resting heart rate, but has no family history of breast cancer or diabetes.

Table 4. Comparison of Test Results for Donor 2

Disease or condition	Profile	Risk predictions			
		Company 1	Company 2	Company 3	Company 4
Alzheimer's disease	Factual	Not tested	Below average	Do not show strong susceptibility	Below average
	Fictitious	Not tested	Not tested	Do not show strong susceptibility	Below average
Atrial fibrillation	Factual	Elevated and typical (research)	Average	Increased susceptibility	Above average
	Fictitious	Elevated and typical (research)	Below average	Average predisposition	Above average
Breast cancer	Factual	Typical	Above average	Average predisposition	Average
	Fictitious	Typical	Average	Average predisposition	Average
Celiac disease	Factual	Elevated and decreased (research)	Below average	Not tested	Below average
	Fictitious	Decreased and elevated (research)	Not tested	Not tested	Below average
Colon cancer	Factual	Typical	Below average	Average predisposition	Below average
	Fictitious	Typical	Average	Increased susceptibility	Below average
Heart attack	Factual	Typical	Average	Average predisposition	Average
	Fictitious	Typical	Below average	Average predisposition	Average
Hypertension	Factual	Typical (research)	Average	Increased susceptibility	Not tested
	Fictitious	Typical (research)	Not tested	Increased susceptibility	Not tested
Leukemia	Factual	Elevated (research)	Average	Increased susceptibility	Not tested

Disease or condition	Profile	Risk predictions			
		Company 1	Company 2	Company 3	Company 4
	Fictitious	Elevated (research)	Not tested	Not tested	Not tested
Multiple sclerosis	Factual	Decreased	Average	Average predisposition	Below average
	Fictitious	Decreased	Not tested	Not tested	Below average
Obesity	Factual	Typical and typical (research)	Average	Average predisposition	About average
	Fictitious	Typical and typical (research)	Not tested	Increased susceptibility	About average
Prostate cancer	Factual	Not applicable	Not applicable	Not applicable	Not applicable
	Fictitious	Not applicable	Not applicable	Not applicable	Not applicable
Restless leg syndrome	Factual	Decreased	Above average	Not tested	About average
	Fictitious	Decreased	Not tested	Not tested	Abut average
Rheumatoid arthritis	Factual	Decreased	Below average	Do not show strong susceptibility	Below average
	Fictitious	Typical	Average	Average predisposition	Below average
Type 1 diabetes	Factual	Elevated	Above average	Do not show strong susceptibility	Not tested
	Fictitious	Elevated	Not tested	Increased susceptibility	Not tested
Type 2 diabetes	Factual	Typical	Average	Average predisposition	About average
	Fictitious	Typical	Above average	Average predisposition	About average

Source: GAO analysis of results from four companies.

Donor 3: Donor 3 is a 48-year-old Caucasian male who has never smoked and rarely drinks. The donor has asthma as well as a family history of heart disease. In Donor 3's fictitious profile, he is a 69-year-old African American male who is overweight, smokes, and is in somewhat poor health. He has a family history of bone and lung cancer, but no history of asthma or heart disease.

Table 5. Comparison of Test Results for Donor 3

Disease or condition	Profile	Risk predictions			
		Company 1	Company 2	Company 3	Company 4
Alzheimer's disease	Factual	Not tested	Average	Increased susceptibility	Above average risk
	Fictitious	Not tested	Not tested	Increased susceptibility	Above average
Atrial fibrillation	Factual	Typical and decreased (research)	Average	Average predisposition	About average
	Fictitious	Typical and decreased (research)	Not tested	Not tested	About average
Breast cancer	Factual	Not applicable	Not applicable	Not applicable	Not applicable
	Fictitious	Not applicable	Not applicable	Not applicable	Not applicable
Celiac disease	Factual	Decreased and typical (research)	Below average	Not tested	Below average
	Fictitious	Decreased and typical (research)	Not tested	Not tested	Below average
Colon cancer	Factual	Typical	Above average	Increased susceptibility	Above average
	Fictitious	Typical	Not tested	Not tested	Above average
Heart attack	Factual	Typical	Average	Average predisposition	Below average
	Fictitious	Typical	Not tested	Not tested	Below average
Hypertension	Factual	Typical (research)	Below average	Increased susceptibility	Not tested
	Fictitious	Typical (research)	Not tested	Not tested	Not tested
Leukemia	Factual	Elevated (research)	Average	Average predisposition	Not tested

Disease or condition	Profile	Risk predictions			
		Company 1	Company 2	Company 3	Company 4
Multiple sclerosis	Fictitious	Elevated (research)	Not tested	Not tested	Not tested
	Factual	Decreased	Average	Average predisposition	Below average
	Fictitious	Decreased	Not tested	Not tested	Below average
Obesity	Factual	Typical and typical (research)	Average	Average predisposition	About average
	Fictitious	Typical and typical (research)	Not tested	Not tested	About average
Prostate cancer	Factual	Typical	Average	Do not show strong susceptibility	Greater than most men's
	Fictitious	Typical	Below average	Average predisposition	Greater than most men's
Restless leg syndrome	Factual	Elevated	Average risk	Not tested	Higher than most people
	Fictitious	Elevated	Not tested	Not tested	Higher than most people
Rheumatoid arthritis	Factual	Elevated	Above average	Average predisposition	Above average
	Fictitious	Elevated	Not tested	Not tested	Above average
Type 1 diabetes	Factual	Elevated	Average	Do not show strong susceptibility	Not tested
	Fictitious	Elevated	Not tested	Not tested	Not tested
Type 2 diabetes	Factual	Typical	Average	Average predisposition	About average
	Fictitious	Typical	Below average	Average predisposition	About average

Source: GAO analysis of results from four companies.

Table 6. Comparison of Test Results for Donor 4

Disease or condition	Profile	Risk predictions			
		Company 1	Company 2	Company 3	Company 4
Alzheimer's disease	Factual	Not tested	Below average	Not tested	Below average
	Fictitious	Not tested	Below average	Average predisposition	Below average
Atrial fibrilation	Factual	Typical and decreased (research)	Below average	Average predisposition	About average
	Fictitious	Typical and decreased (research)	Below average	Average predisposition	About average
Breast cancer	Factual	Not applicable	Not applicable	Not applicable	Not applicable
	Fictitious	Not applicable	Not applicable	Not applicable	Not applicable
Celiac disease	Factual	Elevated and typical (research)	Average	Not tested	Higher risk than most people
	Fictitious	Elevated and typical (research)	Average	Not tested	Higher risk than most people
Colon cancer	Factual	Elevated	Average	Increased susceptibility	Above average
	Fictitious	Elevated	Average	Increased susceptibility	Above average
Heart attack	Factual	Typical	Average	Increased susceptibility	Average
	Fictitious	Typical	Average	Increased susceptibility	Average
Hypertension	Factual	Elevated (research)	Average	Average predisposition	Not tested
	Fictitious	Elevated (research)	Average	Average predisposition	Not tested
Leukemia	Factual	Elevated (research)	Average	Average predisposition	Not tested
	Fictitious	Elevated (research)	Average	Average predisposition	Not tested

Disease or condition	Profile	Risk predictions			
		Company 1	Company 2	Company 3	Company 4
Multiple sclerosis	Factual	Decreased	Average	Average predisposition	Below average
	Fictitious	Decreased	Average	Average predisposition	Below average
Obesity	Factual	Typical and elevated (research)	Average	Average predisposition	About average
	Fictitious	Elevated and typical (research)	Average	Average predisposition	About average
Prostate cancer	Factual	Typical	Above average	Average predisposition	Greater than most men's
	Fictitious	Typical	Above average	Average predisposition	Greater than most men's
Restless leg syndrome	Factual	Decreased	Below average	Not tested	Below average
	Fictitious	Decreased	Below average	Not tested	Below average
Rheumatoid arthritis	Factual	Decreased	Below average	Do not show strong susceptibility	Below average
	Fictitious	Decreased	Below average	Do not show strong susceptibility	Below average
Type 1 diabetes	Factual	Decreased	Average	Do not show strong susceptibility	Not tested
	Fictitious	Decreased	Average	Do not show strong susceptibility	Not tested
Type 2 diabetes	Factual	Typical	Average	Average predisposition	About average
	Fictitious	Typical	Average	Average predisposition	About average

Source: GAO analysis of results from four companies.

Table 7. Comparison of Test Results for Donor 5

Disease or condition	Profile	Risk predictions			
		Company 1	Company 2	Company 3	Company 4
Alzheimer's disease	Factual	Not tested	Above average	Genetic markers are highly correlated with this disease	Above average
	Fictitious	Not tested	Above average	Genetic markers are highly correlated with this disease	Above average
Atrial fibrillation	Factual	Typical and decreased (research)	Below average	Average predisposition	About average
	Fictitious	Typical and decreased (research)	Average	Average predisposition	About average
Breast cancer	Factual	Not applicable	Not applicable	Not applicable	Not applicable
	Fictitious	Not applicable	Not applicable	Not applicable	Not applicable
Celiac disease	Factual	Elevated and decreased (research)	Above average	Not tested	Higher risk than most people
	Fictitious	Elevated and decreased (research)	Above average	Noted tested	Higher risk than most people
Colon cancer	Factual	Decreased	Average	Average predisposition	Average
	Fictitious	Decreased	Average	Average predisposition	Average
Heart attack	Factual	Typical	Above average	Increased susceptibility	Average
	Fictitious	Typical	Above average	Increased susceptibility	Average
Hypertension	Factual	Elevated (research)	Average	Average predisposition	Not tested
	Fictitious	Elevated (research)	Average	Average predisposition	Not tested

Disease or condition	Profile	Risk predictions			
		Company 1	Company 2	Company 3	Company 4
Leukemia	Factual	Elevated (research)	Average	Average predisposition	Not tested
	Fictitious	Elevated (research)	Average	Average predisposition	Not tested
Multiple sclerosis	Factual	Decreased	Average	Average predisposition	Below average
	Fictitious	Decreased	Average	Average predisposition	Below average
Obesity	Factual	Typical and typical (research)	Average	Average predisposition	About average
	Fictitious	Typical and typical (research)	Average	Average predisposition	About average
Prostate cancer	Factual	Typical	Average	Average predisposition	Average
	Fictitious	Typical	Average	Average predisposition	Average
Restless leg syndrome	Factual	Decreased	Above average	Not tested	Higher than most people
	Fictitious	Decreased	Above average	Not tested	Higher than most people
Rheumatoid arthritis	Factual	Decreased	Below average	Do not show strong susceptibility	Below average
	Fictitious	Typical	Below average	Do not show strong susceptibility	Below average
Type 1 diabetes	Factual	Elevated	Average	Average predisposition	Not tested
	Fictitious	Elevated	Average	Average predisposition	Not tested
Type 2 diabetes	Factual	Typical	Average	Average predisposition	Above average
	Fictitious	Elevated	Average	Average predisposition	Above average

Source: GAO analysis of results from four companies.

Donor 4: Donor 4 is a 61-year-old Caucasian male who smokes. The donor has elevated cholesterol, has an elevated resting heart rate, and has had colon cancer. Thirteen years ago, the donor had a pacemaker implanted to treat atrial fibrillation. In Donor 4's fictitious profile, he is a 53-year-old Caucasian male who has never smoked. He has hypertension and prostate cancer but has no family history of colon cancer or atrial fibrillation.

Donor 5: Donor 5 is a 63-year-old Caucasian male who eats a balanced diet and exercises. He has elevated cholesterol and blood sugar. The donor suffers from type 2 diabetes and is obese. He also has a family history of Alzheimer's disease. In Donor 5's fictitious profile, he is a 29-year-old Hispanic male who chews tobacco and suffers from asthma. However, he has no family history of diabetes or Alzheimer's disease.

End Notes

[1] GAO, *Nutrigenetic Testing: Tests Purchased from Four Web Sites Mislead Consumers*, GAO-06-977T (Washington D.C.: July 27, 2006).

[2] The companies are not the same as the companies tested in our 2006 investigation.

[3] The companies also provided consumers with ancestry reports; drug response tests; and predictions for various traits and characteristics, such as eye color. We focused our investigation on testing the companies' disease risk predictions.

[4] Type 1 diabetes is usually first diagnosed in children, teenagers, or young adults. With this form of diabetes, the cells of the pancreas no longer make insulin because the body's immune system has attacked and destroyed them. Type 2 diabetes is the most common form of diabetes. People can develop type 2 diabetes at any age—even during childhood. This form of diabetes usually begins with insulin resistance, a condition in which fat, muscle, and liver cells do not use insulin properly.

[5] In a "research" report contained in the test results, Company 1 also found this donor to be at below-average risk for atrial fibrillation. These conflicting reports will be discussed later in the testimony.

[6] Pauline C. Ng, Sarah S. Murray, Samuel Levy, and Craig J. Venture, *An Agenda for Personalized Medicine, Nature,* vol. 461, October 8, 2009.

[7] Company 3 is the only company that asked consumers to provide medical history information as part of the DNA submission process.

[8] In another report contained in the test results, Company 1 also found this donor to be at average risk for atrial fibrillation. These conflicting reports will be discussed later in the testimony.

[9] In our post-test interviews, Company 1 told us that it is in the process of entering into an agreement with a genetic counseling provider service to which the company will refer interested customers.

[10] An allele is one member of a pair or series of genes that occupy a specific position on a specific chromosome.

In: Direct-to-Consumer Genetic Tests
Editors: T. Hecht and A. F. Maze

ISBN: 978-1-61942- 175-2
© 2012 Nova Science Publishers, Inc

Chapter 2

DIRECT-TO-CONSUMER GENETIC TESTING AND THE CONSEQUENCES TO THE PUBLIC

Jeffrey Shuren

INTRODUCTION

Good morning, I am Dr. Jeff Shuren, Director of the Center for Devices and Radiological Health (CDRH or the Center) at the Food and Drug Administration (FDA or the Agency). I am pleased to be here this morning to explain FDA's recent activities related to direct-to-consumer (DTC) genetic tests and our future plans for the regulation of laboratory-developed tests.

Scientific advances resulting from the Human Genome Project completed in 2003 have expanded our understanding of the genetic contribution to health and disease. These advances have also resulted in the development of new tests that can better identify individuals at risk for particular medical conditions and target medical treatments based on the likelihood that a patient will respond or experience an adverse event based on their individual genetic profile. FDA supports the promise and development of innovative genetic tests.

As Margaret A. Hamburg, M.D., Commissioner of Food and Drugs, and Francis S. Collins, M.D., Ph.D., Director of the National Institutes of Health, note in their jointly authored article entitled "The Path to Personalized Medicine," published in the June 15, 2010, *New England Journal of Medicine*, "Major investments in basic science have created an opportunity for significant progress in clinical medicine. Researchers have discovered hundreds of genes that harbor variations contributing to human illness, identified genetic variability in patients' responses to dozens of treatments, and begun to target the molecular causes of some diseases. In addition, scientists are developing and using diagnostic tests based on genetics or other molecular mechanisms to better predict patients' responses to targeted therapy.... Together, we have been focusing on the best ways to develop new therapies and optimize prescribing by steering patients to the right drug at the right dose at the right time."

However, Dr. Hamburg and Dr. Collins also note that the field of personalized medicine will not make good on that promise if the *in vitro* diagnostic tests on which practitioners and patients rely to inform treatment decisions are inaccurate or the link between what the test measures and its clinical significance is tenuous. Failure to validate the accuracy, reliability, and clinical implications of a test can result in patient harm from misdiagnosis, failure to treat, delay in treatment, inappropriate treatment, or avoidable adverse events.

Overview of Federal Regulation

Congress gave FDA explicit authority to regulate medical devices, including in vitro diagnostic tests, in the 1976 Medical Device Amendments to the Federal Food, Drug, and Cosmetic Act (FD&C Act or the Act). In vitro diagnostic devices (IVDs) are those reagents, instruments, and systems intended for use in the diagnosis of disease or other conditions, including a determination of the state of health, in order to cure, mitigate, treat, or prevent disease or conditions arising from a disease. Genetic tests are a type of IVD.

Under the Act, FDA assigns medical devices to one of three "classes" based upon their attendant risks. The level of regulation applied to IVD devices is based primarily upon risk to the patient of an undetected incorrect test result.

- Class I, subject only to general controls applicable to all devices, is the lowest risk category for a device. Class I IVDs include certain reagents and instruments, as well as a number of highly adjunctive IVD tests, where one test is dependent on the results of another; consequently an incorrect result would generally be detected easily. Most Class I devices are exempt from premarket review. An example

of a Class I test is a luteinizing hormone test that, if it gives a false result, may lead to delayed conception but is unlikely to directly harm the patient.
- Class II, generally subject to general controls and special controls, is the moderate-risk category for a device, and includes many standard laboratory tests, such as chemistry and immunology tests. Most Class II tests are subject to FDA review through premarket notification under section 510(k) of the Act. For example, a false sodium result (a Class II test) may be life-threatening if the error is unrecognized and treatment decisions to correct the sodium level are made based on the false result.
- Class III, subject to premarket approval requirements, is the highest risk category for a device and includes devices and tests that present a potentially unreasonable risk of illness or injury. For example, a false negative result for a hepatitis C virus test (a Class III test) may result in failure to provide appropriate treatment, leading to risk of liver failure due to delayed treatment. In addition, without the knowledge that he or she is infected, the patient may put others at risk by spreading the disease.

Many IVD tests are Class II or Class III devices, and some also may be biological products subject to section 351 of the Public Health Service Act. In addition to premarket controls, the FD&C Act provides FDA with authority to perform post-market review, and monitor adverse events or even mandate a recall if, based on adverse event reports or other data, there is a reasonable probability that a test could cause serious adverse health consequences or death in clinical use.

Federal oversight of IVDs includes oversight of laboratories that perform these tests by the Centers for Medicare and Medicaid Services (CMS) under the Clinical Laboratory Improvement Amendments of 1988 (CLIA), and the Federal Trade Commission (FTC) under the Federal Trade Commission Act (FTCA). Under CLIA, CMS regulates laboratory testing activities performed on humans in the United States for health purposes, covering more than 200,000 laboratory entities. FDA's role under CLIA is to categorize comercially marketed IVDs in terms of their complexity. This complexity categorization determines the stringency of requirements to which the laboratories performing the tests are subject and the attendant personnel education, training, and skill level required.

CLIA and FDA regulations complement one another. CLIA regulations focus on the quality of the clinical testing process, such as laboratory quality control; i.e., daily check that the test is working, external accuracy checks, credentials of laboratory personnel, and documentation of laboratory procedures. FDA regulations address the safety and effectiveness of the diagnostic tests themselves and the quality of the design and manufacture of the diagnostic tests.

Section 5 of the FTCA prohibits unfair or deceptive acts or practices in or affecting commerce. Section 12 of the FTCA specifically prohibits the dissemination of false advertisements for foods, drugs, devices, services, or cosmetics. The FTC analyzes the role of advertising in bringing health-related information to consumers and can bring law enforcement actions against false or deceptive advertising.

Overview of FDA Regulation of Genetic Tests

The purpose of genetic tests includes predicting risk of disease, screening newborns, directing clinical management, identifying carriers, and establishing prenatal or clinical diagnoses or prognoses in individuals, families, or populations. To date, 353 U.S. laboratories have listed themselves on a voluntary website that provides information about laboratories offering genetic tests, but estimates are that there may be as many as 700 laboratories offering such tests.

A genetic test is only subject to FDA oversight if it is a medical device; that is, if it is intended for use in the diagnosis of disease or other conditions, or in the cure, mitigation, treatment, or prevention of disease. For example, a test to determine a person's risk of developing heart disease is a device, whereas a test to determine ancestry is not a device. The type of genetic testing has changed over the past two decades. Whereas early tests tended to identify a single genetic mutation and a patient's risk for developing a disease, some newer tests evaluate thousands of genes or the entire genome and report out risk for a disease based on the combination of dozens of genetic variations.

There currently are two paths to market for a genetic test used in clinical management of patients, as is the case for other IVDs. One is through development of a commercial test kit by an IVD device manufacturer for distribution to multiple laboratories. The Agency has exercised its regulatory authority over these products and has approved several tests for specific genetic factors.

The second pathway is through the development of a test by a laboratory for use only by that laboratory; these are commonly called laboratory-developed tests (LDTs). Conservative estimates are that there are between 2,500 - 5,000 LDTs, including genetic tests that are developed and offered by hundreds of different laboratories.

FDA has the authority to regulate LDTs as it does all IVDs. The extent of FDA oversight of an IVD, such as a genetic test that meets the definition of a device, is based on the risk of an inaccurate result from the test, not who makes the tests or their business models. However, although FDA has the authority to regulate LDTs, FDA has generally exercised enforcement discretion since the device law was passed in 1976. At that time tests made by laboratories were generally low-risk diagnostic tools or relatively simple, well-understood tests that diagnosed rare diseases and conditions, and which were more dependent on expert interpretation. Therefore, the accuracy of the results was more dependent on the expertise of the pathologist/laboratorian than on the design of the test. Furthermore, these LDTs were used by pathologists/laboratorians and the results reported to physicians within a single institution where both professionals were actively involved in the care of the patient being tested. Most genetic tests being offered today are LDTs.

The nature of laboratory-developed tests has changed over the last 30 years, but most dramatically in the last few years. Today, LDTs are increasingly used to assess high-risk but relatively common diseases and conditions, often are used to provide critical information for patient treatment decisions, rely on novel (sometimes preliminary) scientific findings to support their usefulness, often require complex software and may incorporate automated interpretation in lieu of expert interpretation, often are used when there are alternative tests available that have been cleared or approved by FDA, and are performed in commercial laboratory settings that are geographically separate from the patient's primary health care professional and health care setting. In addition, some entities marketed their tests without prior FDA review, claiming that they are LDTs, when they are not. Furthermore, the ability of laboratories to market tests without any regulatory oversight creates a disincentive for traditional manufacturers to develop new tests, thereby stifling innovation.

FDA has observed the following problems with some LDTs in recent years:

- Faulty data analysis
- Exaggerated clinical claims

- Fraudulent data
- Lack of traceability/change control
- Poor clinical study design
- Unacceptable clinical performance

FDA believes that a test used for patient care should have the same assurances of safety and effectiveness whether it is manufactured for distribution to multiple laboratories or created for use in only one laboratory. Premarket review of moderate and high risk LDTs would ensure that the tests are evaluated for analytical validity and clinical validity, based on their claimed intended use, and would provide an independent and unbiased assessment of the data used to support analytical and clinical claims for those LDTs. This is important because when tests are not well validated, the possibility of incorrect results, which can lead to misdiagnosis or inappropriate treatment decisions, increases. Premarket review would also ensure that labeling includes the test claims, the data that support those claims, how the test may be interpreted, and the limitations of the test. FDA's post-market surveillance and enforcement tools ensure that tests remain safe and effective once on the market.

In 2001, the Secretary's Advisory Committee on Genetic Testing recommended that "the Food and Drug Administration should be involved in the review of all new genetic tests regardless of how they are formulated and provided." In 2008, the Secretary's Advisory Committee on Genetics, Health, and Society recommended that FDA address all genetic testing using a risk-based approach.

Historically, FDA's oversight of genetic testing has been focused intensively on commercial test kits. The Agency is now engaging in a public dialogue on how it should develop a consistent, reasonable, and fair approach to all genetic tests, whether packaged as kits or provided as LDTs, to ensure safety and promote innovation.

Genetic Tests being Sold Directly to Consumers

An emerging market segment for the laboratory testing industry is direct-to-consumer testing. A few companies have sought to popularize genetic testing through advertisements and social media. FDA has been aware of these companies marketing to consumers for several years. At the time of the 2006 Government Accountability Office (GAO) investigation of DTC testing, most of these diagnostics were "nutritional genetic" tests—tests to assess what kinds of foods individual consumers should eat and dietary supplements they should take. FDA followed up with the companies and FDA, CDC, and FTC published a cautionary statement on DTC genetic tests.

New DTC genetic tests subsequently came on the market. FDA met with some of these companies starting in 2007. FDA's Center for Devices and Radiological Health, which is responsible for the oversight of these tests, never informed these companies that they could lawfully market their tests without FDA oversight. Instead, the Center met with these companies to have a better understanding of what the companies were in fact doing or planning to do. Initially their business models were not clear and the tests were being marketed for such purposes as "antiquity determinations." However, since then we have seen changes in the number and types of claims being made. For example, one company provided test reports for 17 diseases, conditions, or traits in 2008 but provided over 100 types of results by 2010. In particular, some companies are making claims about high-risk medical indications, such as determining the risk for cancer or the likelihood of responding to a specific drug. Moreover, in many cases the link between the genetic results and the risk of developing a disease or responding/not responding to a drug has not been well-established.

Marketing genetic tests directly to consumers can increase the risk of a test because a patient may make a decision that adversely affects their health, such as stopping or changing the dose of a medication or continuing an unhealthy lifestyle, without the intervention of a learned intermediary. The risk points up the importance of ensuring that consumers are also provided accurate, complete, and understandable information about the limitations of test results they are obtaining.

Recently, we have seen companies more aggressively market directly to consumers. For example, Pathway Genomics corporation (PGC) was poised to offer their Pathway Genomics Genetic Health Report, a home use saliva collection kit, directly to consumers through more than 6,000 Walgreen stores. 23andMe is marketing directly to consumers on Amazon.com.

Although FDA has cleared a number of genetic tests since 2003, none of the genetic tests now offered directly to consumers has undergone premarket review by FDA to ensure that the test results being provided to patients are accurate, reliable, and clinically meaningful.

Because of the escalation in risk and aggressive marketing, FDA notified PGC on May 10, 2010, that their offering appeared to meet the definition of a medical device as that term is defined under the FD&C Act, and clearance or approval by the Agency was necessary in order for them to market their product. The test is intended to report the presence or risk of more than 70 health conditions, including pharmacogenetics (prescription medication response), propensity for complex disease, carrier status, and other information from which one could modify one's lifestyle and health regime, supposedly to live a healthier, longer life. These tests have not been proven safe, effective, or accurate, and patients could be put at risk by making medical decisions based on data that has not received independent premarket review. Following receipt of FDA's letter, PGC stopped marketing directly to consumers.

On June 10, 2010, FDA sent similar letters to four other diagnostic test manufacturing firms that were offering their tests directly to consumers (Knome, Inc; Navigenics; deCODE Genetics; and 23andMe). FDA considers all of these products to meet the statutory definition of a medical device on the basis of the manufacturers' claims about the test results. For example, the tests claimed to describe the genetic basis of specific disease traits or conditions on which consumers may base medical decisions; provide personalized information on which medications are more likely to work given a person's genetic makeup; and provide genetic predispositions for important health conditions and medication sensitivities. In addition, a letter was issued to Illumina, Inc. for supplying unapproved reagents and instrumentation (marked "for research use only" and thus not approved or cleared by FDA) to several DTC manufacturers who use the reagents as critical components in their products being offered directly to consumers for clinical, not research, use. These manufacturers generally have not submitted information on the analytical or clinical validity of their tests to FDA for clearance or approval. All six companies have been invited to discuss the regulatory status of their products further with the Agency. Meetings with the companies are taking place now or have been or are being scheduled. FDA may take additional actions, depending on the outcome of those meetings.

On July 19, 2010, FDA sent similar letters to 15 other firms marketing DTC genetic tests.

Public Meeting on Laboratory-Developed Tests

On July 19 and 20, FDA held a public meeting for the purpose of obtaining input from stakeholders on how the Agency should apply its authority to implement a reasonable, risk-based, and effective regulatory framework for LDTs, including genetic tests, in particular, taking into account circumstances unique to LDTs and to avoid any duplication with CLIA. We provided an overview of the history and current regulatory status of LDTs. The meeting discussions focused on:

- Patient Considerations
- Challenges for Laboratories
- Direct-to-Consumer Marketing of Testing
- Education and Outreach

Each session consisted of presentations from interested stakeholders, followed by an expert panel discussion and question-and-answer period. The meeting record will be held open for an additional comment period of 60 days, after which FDA will collect and review all comments and information presented. Subsequently, the Agency will move forward in developing a draft oversight framework for public comment as quickly as possible. FDA intends to phase in over time whatever framework it constructs, based on the level of risk of the test.

CONCLUSION

Mr. Chairman, I commend the Subcommittee's efforts to inform the ongoing dialog about the safety and accuracy of genetic tests being marketed today. FDA is working toward a reasonable and fair approach to regulation that can give patients and doctors confidence in these tests and facilitate progress in personalized medicine. Mr. Chairman, that concludes my formal remarks. I will be pleased to answer any questions the Subcommittee may have.

In: Direct-to-Consumer Genetic Tests
Editors: T. Hecht and A. F. Maze

ISBN: 978-1-61942-175-2
© 2012 Nova Science Publishers, Inc

Chapter 3

TESTIMONY OF JAMES P. EVANS MD, PH.D, HEARING OF THE HOUSE ENERGY AND COMMERCE COMMITTEE'S SUBCOMMITTEE ON OVERSIGHT AND INVESTIGATIONS, JULY 20, 2010

James P. Evans

Thank you very much for inviting me to testify. I am a physician and scientist who specializes in medical genetics. My research involves the use of emerging technologies to analyze the human genome for genes involved in cancer predisposition and the ways in which people use genetic information. I am the Editor-in-Chief of *Genetics in Medicine*, the official journal of the American College of Medical Genetics. But first and foremost I am a physician. I am a board certified internist who has a general medical practice. I am also board certified in Clinical Medical Genetics and in Molecular Genetic Diagnostics in which capacity I see and test patients who have, or are at risk of having, genetic disorders such as predisposition to cancer.

The breathtaking pace of discovery in the field of genetics is providing new opportunities for rapidly and inexpensively analyzing the human genome. We are now able to routinely query an individual's genome at over 1 million sites and the "$1,000 genome", in which access to one's entire genetic code will be feasible for many individuals, will soon be a reality.

Such advances in technology have great promise to revolutionize medicine and usher in a new era of genomic medicine. These advances will lead to great progress in our basic understanding of disease, improved diagnostic abilities, new therapies and personalized prescription of drugs.

But the rapid pace of technological progress has left us understandably impatient for immediate application to patient care. Like scientists and doctors, the public is curious and hopeful about genetics and has demonstrated an interest in analyzing and understanding their own genome. Indeed, we may be approaching an era in which much, if not most, genetic testing could be done outside the confines of the traditional doctor's office or medical setting.

In part to meet this burgeoning interest, a wide range of direct to consumer (DTC) genetic testing entities has arisen, a potentially positive development for both patients and the public. We should encourage individuals to be involved in, and be the primary directors of, their own healthcare. Truly participatory, individualized medicine is a worthy goal and one we should strive for. People should have access to the information contained in their own genome.

But it is also critical that they be assured that the information they receive is of high quality, that they have recourse to disinterested advice about the meaning of that information, that their privacy be protected and that claims made by the purveyors of such testing comport with reality.

Unfortunately this is not always the case at present. The most egregious problem – and the most remediable – is the distinct gap between claims made by the providers of such services and the value of the information actually imparted. Most of the purveyors of DTC genetic testing appeal both implicitly and explicitly to the purported medical value of the genetic tests in question. We hear claims that scanning your genome for genetic variants provides a "road map to better health", allows one to "take control of your health future" or that "knowledge is power" with regard to disease. Indeed, these are the central advertising logos of the three most prominent players in the genomic DTC arena. Yet on each page of every report provided to patients by these companies, some variant of the following disclaimer is made: "Information provided is not intended as, nor does it provide, medical advice, treatment, diagnosis, or treatment guidelines." The explicit health claims and the accompanying disclaimer (in tiny font) cannot both be true. And indeed they are not. The disclaimer is correct. Such information, by and large, utterly lacks medical significance. This would be true even if we understood how to interpret such tests, which, as clearly demonstrated by scrutiny of the literature and the recent GAO investigation, we do not.

It is often submitted by boosters of such technology that mere knowledge of one's risks will be of benefit to an individual. Yet, little evidence suggests that this is the case. Statistics about risk are tricky. I know, to a first approximation, what you, the reader of this document, will likely die of... cardiovascular disease or cancer. These maladies are not called "common diseases" for nothing. They are exceedingly common and one is at considerable risk for them regardless of whether one happens to be at a *relatively* increased or decreased risk when compared with the average individual in the population. Thus, even for those at decreased relative risk, the chances are that they too will die of one of these common diseases. Thus, finding out that you're at double or half the "average" risk of a common disease is simply not medically meaningful.

Likewise, for rare diseases, what does defining your risk really mean? The risk of a US citizen developing Crohns disease, a disease of the GI tract, is about 1/1000. In what way is it useful to know that I'm instead at a 1/500 risk or a 1/2000 risk?

It is instructive to examine how we use risk information in pursuit of better health. Your doctor doesn't measure your cholesterol and blood pressure because simple knowledge of that risk information is beneficial to your health. Rather, she measures it because we have ways of altering your blood pressure or cholesterol. As a physician, I simply don't know what to do with the knowledge that I or my patient is at, say, a 40% increased risk for prostate cancer. We have no interventions that make that information useful.

Some claim that knowledge of an increased risk will motivate people to live more healthy life styles. Yet there is no good data thus far that genetic information has any special qualities that will motivate individuals any more effectively than do our current admonitions.

But what if I'm wrong? What if there really is something inherently special about genetic information that will induce behavioral change? I sincerely doubt that this will be the case but let's grant that dubious proposition for a minute. If so, we have an even bigger problem. Because for everyone I find to be at increased risk of, say, heart disease, I am mathematically guaranteed to find another at decreased risk. If genetic information has magical abilities to affect behavior then we run the inevitable risk that such information will induce adverse behavior in the other half of the population, to their ultimate detriment. The bottom line is that whether you are at increased risk or decreased risk of disease, a healthy lifestyle will benefit you and there is little to be gained from finely parsing that risk. The gap between claims and reality should be closed. And it doesn't even require new regulations, just enforcement of existing standards that are, at least in part, promulgated and promoted by the FTC.

Another important issue before us is what sources of information the public has about the meaning of their results. I would argue that the vast majority of test results provided by most DTC genetic companies are simply of entertainment, not medical value. As such I see little potential for harm and see no problem with the public having full access to such information - as long as it is not oversold in the way I've just been describing.

But mixed in with trivial and fun tests (that, for example, assay your likelihood of having thick earwax or liking Brussels sprouts) are a few tests offered by such companies that have very serious medical consequences. For example, one major purveyor's panel of DTC tests include, along with trivial matters, a test for specific mutations which result in an exceedingly high risk of breast and ovarian cancer. Thus, having signed up for innocuous information about one's ancestry and possible food preferences, women may also find out via the company's website that they should perhaps consider bilateral mastectomies and removal of their ovaries. Startlingly, the recipients of such information have no recourse to even talk with a qualified professional about their result and its implications. I think that people should be free to get medical tests on their own terms. But if one takes on the responsibility of informing someone that they have tested positive for a mutation that could well lead to very serious – indeed life changing - consequences, then one should ensure that the individual can at least pick up a phone and talk with someone knowledgeable about its implications for them and for their loved ones. I don't leave my patients in the lurch when I discover devastating information about their health and doing so should not become a new standard of the internet age.

Ensuring quality testing is also of paramount importance as we try to realize the potential of genomic information. Simply put, if such information has true medical value, then its quality should be ensured like any other medical test. This is not too much to ask. I applaud the recent move of the FDA to take a risk-calibrated approach to the regulation of such testing. Their action is especially timely given the recent mix up of 87 samples which occurred from a major purveyor of DTC genomics. In formulating appropriate regulations it is important to keep in mind that risk calibration is possible. There is no reason that each test must be regulated to the same degree. Rather, the seriousness of the implications of a given test can guide the degree to which it must be regulated.

Protecting the public's privacy is critical. A tiny sample of your DNA can serve to differentiate you from every other human who has ever lived. Thus, it seems reasonable that the public should be assured that their samples and their genomic information are protected. What do we do when a company goes bankrupt and ownership of your uniquely identifying genetic information suddenly may become the property of a venture capital firm? We need clear and enforceable guidelines for how such information is handled by its (likely numerous) owners.

As we seek to employ genomic information in healthcare it's critical and sometimes difficult to remember that medicine and science are very different pursuits. Unfortunately good ideas are insufficient to guide the practice of medicine. We've learned that we must demand evidence of efficacy and safety before we translate what seem like good ideas into medical care. If we do not it is our patients who will inevitably pay the price.

No one is more excited about the future of genomics than I am, nor feels more strongly that it has the potential to usher in a new era of medicine that will benefit us all. I welcome the entry of quality-minded and responsible entrepreneurs into the field. Medicine is often validly criticized for being too slow to change and I think we have plenty to learn from innovative cutting edge companies, some of the representatives of which are also testifying today.

But as a physician who deals these issues daily I do not feel that it is paternalistic to ask that the public not be deceived by exaggerated claims, that their privacy be protected, that tests be of high quality and that they have recourse to unbiased information about the meaning of their results. Regulation does not mean proscription. We can embrace an exciting future in which the public has access to its genomic information but we should do so in a responsible manner and risk-calibrated regulation is part of the answer. Indeed, it seems obvious to me that the interests of companies and the public are actually fully aligned since both their long-term business interests and public's health will thrive only when tests and the claims made for those tests are trusted.

I believe that the public deserves access to the information contained in their own genomes. But they also deserve an honest accounting of what such information means and the assurance that it is derived in a manner that ensures quality, reliability and confidentiality.

James P. Evans MD, Ph.D
Bryson Distinguished Professor of Genetics and Medicine

In: Direct-to-Consumer Genetic Tests
Editors: T. Hecht and A. F. Maze

ISBN: 978-1-61942-175-2
© 2012 Nova Science Publishers, Inc

Chapter 4

GENETIC TESTING: SCIENTIFIC BACKGROUND FOR POLICYMAKERS

Amanda K. Sarata

SUMMARY

Congress has considered, at various points in time, numerous pieces of legislation that relate to genetic and genomic technology and testing. These include bills addressing genetic discrimination in health insurance and employment, personalized medicine, the patenting of genetic material, and the quality of clinical laboratory tests, including genetic tests. The focus on these issues signals the growing importance of the public policy issues surrounding the clinical and public health implications of new genetic technology. As genetic technologies proliferate and are increasingly used to guide clinical treatment, these public policy issues are likely to continue to garner considerable attention. Understanding the basic scientific concepts underlying genetics and genetic testing may help facilitate the development of more effective public policy in this area.

Most diseases have a genetic component. Some diseases such as Huntington's Disease are caused by a specific gene. Other diseases, such as heart disease and cancer, are caused by a complex combination of genetic and environmental factors. For this reason, the public health burden of genetic disease is substantial, as is its clinical significance. Experts note that society has recently entered a transition period in which specific genetic knowledge is becoming critical to the delivery of effective health care for everyone. Therefore, the value of and role for genetic testing in clinical medicine is likely to increase significantly in the future.

INTRODUCTION

Virtually all disease has a genetic component.[1] The term "genetic disease" has traditionally been used to refer to rare monogenic (caused by a single gene) inherited disease, for example, cystic fibrosis. However, we now know that many common complex human diseases, including common chronic conditions such as cancer, heart disease and diabetes, are influenced by several genetic and environmental factors.[2] For this reason, they could all be said to be "genetic diseases." Considering this broader definition of genetic disease, the public health burden of genetic disease can be seen to be substantial. In addition, an individual patient's genetic make-up, and the genetic make-up of his disease, will help guide clinical decision making. Experts note that "(w)e have recently entered a transition period in which specific genetic knowledge is becoming critical to the delivery of effective health care for everyone."[3] This sentiment is still broadly shared, despite the fact that the translation to practice has perhaps been slower than anticipated due to the lack of an evidence base to inform clinical validity and utility determinations for many genomic technologies. The value of and role for genetic testing in clinical medicine is likely to increase significantly in the future. As the role of genetics in clinical medicine and public health continues to grow, so will the importance of public policy issues raised by genetic technologies.

Science is only beginning to unlock the complex nature of the interaction between genes and the environment in common disease, and their respective contributions to the disease process. The information gleaned from the Human Genome Project will help, and is currently helping, scientists and clinicians to identify common genetic variation that contributes to disease, primarily through genome-wide association studies (GWAS).[4] However, researchers have identified a significant translational gap between genetic discoveries and application in clinical and public health practice and note that "the pace of implementation of genome-based applications in health care and population health has been slow."[5] Efforts are underway to close this gap and expedite translation into practice, specifically the recent development of the NIHCDC collaborative Genomic Applications in Practice and Prevention Network. Experts note that the moderate effect of many common variants, uncovered by

GWAS, has helped to underscore the multifactorial etiology of complex disease, and that substantially greater research efforts will be required to detect "missing" genetic influences.[6] In addition, research conducted utilizing large population databases that collect health, genetic, and environmental information about entire populations will likely provide more information about the genetic and environmental underpinnings of common diseases. Many countries have established such databases, including Iceland, the United Kingdom, and Estonia. The knowledge of the potential relevance of genetic information to the clinical management of nearly all patients coupled with the lack of complete information about the genetic and environmental factors underlying disease creates a challenging climate for public policymaking.

In many cases, the results of genetic testing may be used to guide clinical management of patients and a particularly prominent role is anticipated in the realm of preventive medicine.[7] For example, more frequent screening may be recommended for individuals at increased risk of certain diseases by virtue of their genetic make-up, such as colorectal and breast cancer. In some cases, prophylactic surgery may even be indicated. Decisions about courses of treatment and dosing may also be guided by genetic testing, as might reproductive decisions (both clinical and personal). However, many diseases do not have any treatment available (for example, Huntington's Disease). In these cases, the benefits of genetic testing lie largely in the information they provide an individual about his or her risk of future disease or current disease status. The value of genetic information in these cases is personal to individuals, who may choose to utilize this information to help guide medical and other life decisions for themselves and their families. The information can affect decisions about reproduction, the types or amount of health, life, or disability insurance to purchase, or career and education choices. As genetic research continues to advance rapidly, it will often be the case that genetic testing may be able to provide information about the probability of a health outcome without an accompanying treatment option. This situation again creates unique public policy challenges, for example, in terms of the financing of genetic testing services and education about the value of testing.

Concerns about privacy and the use and misuse of genetic information, as well as issues of genetic exceptionalism[8] and genetic determinism[9], may need to be balanced with the potential of genetics and genetic technology to change how care is delivered and to personalize medical care and treatment of disease.

This report will summarize basic scientific concepts in genetics and will provide an overview of genetic tests, their main characteristics, and the key policy issues they raise.

FUNDAMENTAL CONCEPTS IN GENETICS

The following section explains key concepts in genetics that are essential for understanding genetic testing and issues associated with testing that are of interest to Congress.

Cells Contain Chromosomes

Humans have 23 pairs of chromosomes in the nucleus of most cells in their bodies. These include 22 pairs of autosomal chromosomes (numbered 1 through 22) and one pair of sex chromosomes (X and Y). One copy of each autosomal chromosome is inherited from the mother and from the father, and each parent contributes one sex chromosome.

Many syndromes involving abnormal human development result from abnormal numbers of chromosomes (such as Down Syndrome). Other diseases, such as leukemia, can be caused by breaks in or rearrangements of chromosome pieces.

Chromosomes Contain DNA

Chromosomes are composed of deoxyribonucleic acid (DNA) and protein. DNA is comprised of complex chemical substances called bases. Strands made up of combinations of the four bases (adenine (A), guanine (G), cytosine (C) and thymine (T)) twist together to form a double helix (like a spiral staircase). Chromosomes contain almost 3 billion base pairs of DNA that code for about 20,000-25,000 genes (this is a current estimate, although it may change and has changed several times since the publication of the human genome sequence).[10]

DNA Codes for Protein

Proteins are fundamental components of all living cells. They include enzymes, structural elements, and hormones. Each protein is made up of a specific sequence of amino acids. This sequence of amino acids is determined by the specific order of bases in a section of DNA. A gene is the section of DNA which contains the sequence which corresponds to a specific protein.

Changes to the DNA sequence, called mutations, can change the amino acid sequence. Thus, variations in DNA sequence can manifest as variations in the protein which may affect the function of the protein. This may result in, or contribute to the development of, a genetic disease.

Genotype Influences Phenotype

Though most of the genome is very similar between individuals, there can be significant variation in physical appearance or function between individuals. In other words, although we share most of the genetic material we have, we can see that there are significant differences in our physical appearance (height, weight, eye color, etc.). Humans inherit one copy (or allele) of most genes from each parent. The specific alleles that are present on a chromosome pair constitute a person's genotype. The actual observable, or measurable, physical trait is known as the phenotype. For example, having two brown-eye color alleles would be an example of a genotype and having brown eyes would be the phenotype.

Many complex factors affect how a genotype (DNA) translates to a phenotype (observable trait) in ways that are not yet clear for many traits or conditions. Study of a person's genotype may determine if a person has a mutation associated with a disease, but only observation of the phenotype can determine if that person actually has physical characteristics or symptoms of the disease. Generally, the risk of developing a disease caused by a single mutation can be more easily predicted than the risk of developing a complex disease caused by multiple mutations in multiple genes and environmental factors. Complex diseases, such as heart disease, cancer, immune disorders, or mental illness, for example, have both inherited and environmental components that are very difficult to separate. Thus, it can be difficult to determine whether an individual will develop symptoms, how severe the symptoms may be, or when they may appear.

GENETIC TESTS

What is a Genetic Test?

Scientifically, a genetic test may be defined as:

an analysis performed on human DNA, RNA, genes, and/or chromosomes to detect heritable or acquired genotypes, mutations, phenotypes, or karyotypes that cause or are likely to cause a specific disease or condition. A genetic test also is the analysis of human proteins and certain metabolites, which are predominantly used to detect heritable or acquired genotypes, mutations, or phenotypes.[11]

Once the sequence of a gene is known, looking for specific changes is relatively straightforward using the modern techniques of molecular biology. In fact, these methods have become so advanced that hundreds or thousands of genetic variations can be detected simultaneously using a technology called a microarray.

Policy Issues

The way genetic test is defined can be very important to the development of genetics-related public policy. For example, the above scientific definition is very broad, including both predictive and diagnostic tests and analyses on a broad range of material (nucleic acid, protein, and metabolites), but this may not be the best way to achieve certain policy goals. It may sometimes be desirable to limit the definition only to predictive, and not diagnostic, genetic testing because often, predictive tests raise public policy concerns that diagnostic tests do not (see "What are the different Types of Genetic Tests?"). In other cases, it may be desirable to limit the definition to only analysis of specific material, such as DNA, RNA, and chromosomes, but not metabolites or proteins, for example, to help avoid capturing certain types of tests, such as newborn screening tests, in the scope of a proposed law. Policies extending protection against discrimination may aim to be as broad as possible, whereas policies addressing the stringency of oversight of genetic tests may aim to be more limited (to predictive or probabilistic tests only, or to those for conditions with no treatment, for example).

How Many Genetic Tests are Available?

As of January 2010, genetic tests are available for 1,889 diseases. Of those tests, 1,626 are available for clinical diagnosis, while 263 are available for research only.[12] The majority of these tests are for single-gene rare diseases. Asked about the realistic promise of genetic technology, Francis Collins, the Director of the National Human Genome Research Institute predicted,

I think we can count on the availability within the next decade of a panel of genetic tests that are going to be offered to all of us to determine our risk of common illnesses, focused particularly on those diseases for which there is some intervention available for those found to be at high risk.[13]

What are the Different Types of Genetic Tests?

Most clinical genetic tests are for rare disorders, but increasingly, tests are becoming available to determine susceptibility to common, complex diseases and to predict response to medication.

With respect to health-related tests (i.e., excluding tests used for forensic purposes, such as "DNA fingerprinting"), there are two general types of genetic testing: diagnostic and predictive. Genetic tests can be utilized to identify the presence or absence of a disease (diagnostic). Predictive genetic tests can be used to predict if an individual will definitely get a disease in the future (predictive-presymptomatic) or to predict the risk of an individual getting a disease in the future (predictive-predispositional). For example, testing for mutations in the BRCA1 and/or BRCA2 genes provides probabilistic information about how likely an individual is to develop breast cancer in his or her lifetime (predispositional). The genetic test for Huntington's Disease provides genetic information that is predictive in that it allows a physician to predict with certainty whether an individual will develop the disease, but does not allow the physician to determine when the onset of symptoms will actually occur (presymptomatic). In both of these examples, the individual does not have the clinical disease at the time of genetic testing, as they would with diagnostic genetic testing.

Within this broader framework of diagnostic and predictive genetic tests, several distinct types of genetic testing can be considered. Reproductive genetic testing can identify carriers of genetic disorders, establish prenatal diagnoses or prognoses, or identify genetic variation in embryos before they are used in in vitro fertilization. Reproductive testing, such as prenatal testing, may be either diagnostic or predictive in nature. Newborn screening is a type of genetic testing that identifies newborns with certain metabolic or inherited conditions (although not all newborn screening tests are genetic tests). Traditionally, most newborn screening has been diagnostic, but recently several states have added some predictive genetic testing to their panels of newborn screening (for example, Maryland includes testing for cystic fibrosis).[14]

Finally, pharmacogenomic testing, or testing to determine a patient's likely response to a medication, may be considered either diagnostic or predictive, depending on the context in which it is being utilized (i.e., before administration of a medication to determine potential effectiveness, dosing levels, or potential adverse interactions or events vs. after administration and manifestation of a clinical event, for use in determining the basis of the specific event or outcome in the particular patient).

Policy Issues

Generally, predictive genetic testing (both presymptomatic and predispositional), rather than diagnostic testing, raises more complex ethical, legal and social issues. For example, issues surrounding insurance coverage and reimbursement for this type of test, especially if no treatment is available, are far more complex than with diagnostic genetic testing. A private insurer may feel that paying for a test that predicts the onset of a disease with no treatment is not cost-effective. Even more complicated are cases where the test only shows an increased probability of getting a disease.

Another issue is the oversight of genetic tests. Strong oversight of genetic tests may be more important where the information is probabilistic rather than diagnostic and when a treatment is not available. Additionally, stronger regulation of direct-to-consumer marketing of genetic tests, or direct access testing, may be desirable in cases where a test is probabilistic rather than diagnostic.

Finally, issues of genetic discrimination may be different with predictive testing than they are with diagnostic testing. Genetic discrimination may be defined as differential treatment in either health insurance coverage or employment based upon an individual's genotype. Discriminatory action based on the possibility of something happening in the future, or even the certainty of it happening in the future, might raise more concern than would action taken based upon diagnostic information. With probabilistic genetic information, the health outcome of concern may never manifest, or if it is certain to, may not manifest for decades into the future.

The Genetic Test Result

Genetic tests can provide information about both inherited genetic variations, that is, the individual's genes that were inherited from their mother and father, as well as about acquired genetic variations, such as those that cause some tumors. Acquired mutations are not inherited, but rather are acquired in DNA due to replication errors or exposure to mutagenic chemicals and radiation (e.g., UV rays).

DNA-based testing of inherited genetic variants differs from other medical testing in important ways: it can have exceptionally long-range predictive powers over the lifespan of an individual; it can predict disease or increased risk for disease in the absence of clinical signs or symptoms; it can reveal the sharing of genetic variants within families at precise and calculable rates; and, at least theoretically, it has the potential to generate a unique identifier profile for individuals. Also, unlike most other medical tests, the stability of DNA means that most genetic tests can be performed on material from a body and continue to provide information after the individual has died.

Genetic changes to inherited genes can be acquired throughout a person's life. Tests that are performed for acquired genetic markers that occur with a disease have implications only for individuals with the disease, and not family members. Tests for acquired genetic changes are also usually diagnostic rather than predictive, since these tests are generally performed after symptoms present.

Pharmacogenomic testing may be used to determine acquired genetic variations in disease tissue (i.e., acquired variations in a tumor) or may be used to determine inherited variations in an individual's drug metabolizing enzymes. For example, with respect to determining acquired variation in disease tissue, a tumor may have acquired genetic changes that make it different from normal tissue that may also render that tumor susceptible or resistant to chemotherapy. With respect to inherited variation in drug metabolizing enzymes, an individual may be found to be a slow metabolizer of a certain type of drug (statins, for example) and this information can be used to guide both drug choice and dosing.

Policy Issues

In some cases, people feel differently about their genetic information than they do about other medical information, a sentiment embodied by the concept of genetic exceptionalism. This may be based on the stated differences between genetic testing and other medical testing, but also may be based on personal belief that genetic information is powerful and different than other medical information. For this reason, public policies addressing genetic discrimination in health insurance, employment and sometimes life insurance proliferated at the state level in the 1990s, and the Genetic Information Nondiscrimination Act of 2008 (P.L. 110-233) was signed into law on May

21, 2008. Whether genetic information is somehow different from other medical information; whether it can be separated logically from other medical information; and whether it deserves special protection are all important public policy issues.

Pharmacogenomic testing is important because it will help provide the foundation for personalized medicine. Personalized medicine is healthcare based on individualized diagnosis and treatment for each patient determined by information at the genomic level. Many public policy issues are associated with personalized medicine. For example, there is some uncertainty currently as to how health insurers will assess and choose to cover pharmacogenomic testing as it becomes available. In addition, there are issues surrounding the regulation of pharmacogenomic testing and the encouragement of the co-development of drugs and diagnostic genetic tests (companion diagnostics).

Characteristics of Genetic Tests

Genetic tests function in two environments: the laboratory and the clinic. Genetic tests are evaluated based primarily on three characteristics: analytical validity, clinical validity, and clinical utility.

Analytical Validity. Analytical validity is defined as the ability of a test to detect or measure the analyte it is intended to detect or measure.[15] This characteristic is critical for all clinical laboratory testing, not only genetic testing, as it provides information about the ability of the test to perform reliably at its most basic level. This characteristic is relevant to how well a test performs in a laboratory.

Clinical Validity. The clinical validity of a genetic test is its ability to accurately diagnose or predict the risk of a particular clinical outcome. A genetic test's clinical validity relies on an established connection between the DNA variant being tested for and a specific health outcome. Clinical validity is a measure of how well a test performs in a clinical rather than laboratory setting. Many measures are used to assess clinical validity, but the two of key importance are clinical sensitivity and positive predictive value. Genetic tests can be either diagnostic or predictive and, therefore, the measures used to assess the clinical validity of a genetic test must take this into consideration. For the purposes of a genetic test, positive predictive value can be defined as the probability that a person with a positive test result (i.e., the DNA variant tested for is present) either has or will develop the disease the test is designed to detect. Positive predictive value is the test measure most commonly used by physicians to gauge the usefulness of a test to clinical management of patients. Determining the positive predictive value of a predictive genetic test may be difficult because there are many different DNA variants and environmental modifiers that may affect the development of a disease. In other words, a DNA variant may have a known association with a specific health outcome, but it may not always be causal. Clinical sensitivity may be defined as the probability that people who have, or will develop a disease, are detected by the test.

Clinical Utility. Clinical utility takes into account the impact and usefulness of the test results to the individual and family and primarily considers the implications that the test results have for health outcomes (for example, is treatment or preventive care available for the disease). It also includes the utility of the test more broadly for society, and can encompass considerations of the psychological, social and economic consequences of testing.

Policy Issues

These three characteristics of genetic tests have important ties to public policy issues. For example, although the analytical validity of genetic tests is regulated by the Centers for Medicare and Medicaid Services (CMS) through the Clinical Laboratory Improvement Amendments (CLIA) of 1988 (P.L. 100-578), the clinical validity of the majority of genetic tests is not regulated at all. This has raised concerns about direct-to-consumer marketing of genetic tests where the connection between a DNA variant and a clinical outcome has not been clearly established. Marketing of such tests to consumers directly may mislead consumers into believing that the advice given them based on the results of such tests could improve their health status/outcomes when in fact there is no scientific basis underlying such an assertion. This issue was the subject of a July 2006 hearing by the Senate Special Committee on Aging. In addition, clinical utility and clinical validity both figure prominently into coverage decisions by payers, but a lack of data often hinders coverage decisions, potentially leaving patients without coverage for these tests.

Coverage by Health Insurers

Health insurers are playing an increasingly large role in determining which medical tests are available by deciding which tests they will pay for as part of patient benefit packages. Many aspects of genetic tests, including their clinical validity and utility, may complicate the coverage decision-making process for insurers. While insurers require that a test be approved by the Food and Drug Administration (when required), they also want evidence that it is "medically necessary;" that is, evidence demonstrating that a test will affect a patient's health outcome in a positive way. This additional requirement of evidence of improved health outcomes underscores the importance of patient participation in long-term research in genetic medicine. Particularly for genetic tests, data on health outcomes may take a very long time to collect.

Policy Issues

Decisions by insurers to cover new genetic tests have a significant impact on the utilization of such tests and their eventual integration into the health care system. The integration of personalized medicine into the health care system will be significantly affected by coverage decisions. Although insurers are beginning to cover pharmacogenomic tests and treatments, the high cost of such tests and treatments often means that insurers require very stringent evidence that they will improve health outcomes. In addition, coverage for many genetic tests and services, which may be considered preventive and therefore would have to undergo special review by the USPSTF and be authorized for coverage by the Secretary of HHS, may not be granted under Medicare.[16] As Medicare coverage decisions are often looked to by private insurers as a guide for their own coverage decisions, it is difficult to predict what effect this might have on the uptake and utilization of genetic tests more broadly.

Regulation of Genetic Tests by the Federal Government

Genetic tests are regulated by the FDA and CMS through CLIA. FDA regulates genetic tests that are manufactured by industry and sold for clinical diagnostic use. These test kits usually come prepackaged with all of the reagents and instructions that a laboratory needs to perform the test and are considered to be products by the FDA. FDA requires manufacturers of the kits to make sure that the test detects what they say it will, in the patient population in which they intend the test to be used. With respect to the characteristics of a genetic test, this process requires manufacturers to prove that their test is clinically valid. Depending on the perceived risk associated with the intended use promoted by the manufacturer, genetic tests must either prove that they are safe and effective, or that they are substantially equivalent to something that is already on the market that has the same intended use.

Most genetic tests are performed, not with test kits, but rather as laboratory testing services (or "homebrew" tests), meaning that clinical laboratories themselves perform the test in-house and make most or all of the reagents used in the tests. Homebrew tests are not currently regulated by the FDA in the way kits are and, therefore, the clinical validity of the vast majority of genetic tests is not regulated. The FDA does regulate certain components used in homebrew tests, known as Analyte Specific Reagents (ASRs), if the ASR is commercially available. If the ASR is made in-house by a laboratory performing a homebrew test, the test is not regulated at all by the FDA. This type of test is known as a "homebrew-homebrew" test.

Any clinical test that is performed with results returned to the patient must be performed in a CLIA-certified laboratory. CLIA is primarily administered by CMS in conjunction with the Centers for Disease Control and Prevention (CDC) and the FDA.[17] FDA determines the category of complexity of the test so that laboratories know which parts of CLIA they must follow. As previously noted, CLIA regulates the analytical validity of a clinical laboratory test only. It generally establishes requirements for laboratory processes, such as personnel training and quality control/quality assurance programs. CLIA requires laboratories to prove that their tests work properly, to maintain the appropriate documentation, and to show that tests are interpreted by laboratory professionals with the appropriate training. However, CLIA does not require that tests made by laboratories undergo any review by an outside agency to see if they work properly. Supporters of the CLIA regulatory process argue that regulation of the testing process gives the laboratories optimal flexibility to modify tests as new information becomes available. Critics argue that CLIA does not go far enough to assure the accuracy of genetic tests since it only addresses analytical validity and not clinical validity.

End Notes

[1] Collins, F.S. and V.A. McCusick. (2001) "Implications of the Human Genome Project for Medical Science." *Journal of the American Medical Association* 285:540-544.

[2] Manolio, T.A. et al. (2009) "Finding the missing heritability of complex diseases." *Nature* 461(8): 747-753.
[3] Guttmacher, A.E. and F.S. Collins. (2002) "Genomic Medicine - A Primer." *New England Journal of Medicine* 347(19): 1512-1520.
[4] Genome-wide association studies are defined by the National Human Genome Research Institute as "...an approach used in genetics research to associate specific genetic variations with particular diseases. The method involves scanning the genomes from many different people and looking for genetic markers that can be used to predict the presence of a disease." http://www.genome.gov/glossary/index.cfm?id=91
[5] Khoury M.J. et al. (2009) "The Genomic Applications in Practice and Prevention Network." *Genetics in Medicine* 11(7): 488-494.
[6] See note 2 at page 751.
[7] Collins F.S. (2010) "Opportunities for Research and NIH." *Science* 327: 36-37.
[8] *Genetic exceptionalism* is the concept that genetic information is inherently unique, should receive special consideration, and should be treated differently from other medical information.
[9] *Genetic determinism* is the concept that our genes are our destiny and that they solely determine our behavioral and physical characteristics. This concept has mostly fallen out of favor as the substantial role of the environment in determining characteristics has been recognized.
[10] National Research Council, Reaping the Benefits of Genomic and Proteomic Research: Intellectual Property Rights, Innovation, and Public Health. Washington, DC: National Academies Press (2006); p. 19.
[11] Report of the Secretary's Advisory Committee on Genetic Testing (SACGT), "*Enhancing the Oversight of Genetic Tests: Recommendations of the SACGT,* " July 2000, at http://oba.od.nih.gov/oba/sacgt/reports/oversight_report.pdf.
[12] See http://www.genetests.org for information on disease reviews, an international directory of genetic testing laboratories, an international directory of genetics and prenatal diagnosis clinics, and a glossary of medical genetics terms.
[13] E. Rabinowitz, "Genetics in Medicine: Hype or Real Promise?" *Health plan*, January/February 2003.
[14] Newborn Screening Programs, Family Health Administration, Maryland Department of Health and Mental Hygiene. http://www.fha.state.md.us/genetics/nbs_bloodspot.cfm.
[15] An analyte is defined as a substance or chemical constituent undergoing analysis.
[16] For more information on Medicare preventive services generally, please see CRS Report R40978, *Medicare Coverage of Clinical Preventive Services*, by Sarah A. Lister and Kirsten J. Colello.
[17] See http://www.cms.hhs.gov/CLIA/.

CHAPTER SOURCES

The following chapters have been previously published:

Chapter 1 is an edited, reformatted and augmented version of a United States Government Accountability Office publication, GAO-10-847T, dated July 22nd, 2010.

Chapter 2 is an edited, reformatted and augmented version of the statement of Jeffrey Shuren, M. D., before the Subcommittee on Oversight and Investigations Committee on Energy and Commerce, U.S House of Representatives, dated July 22nd, 2010.

Chapter 3 is an edited, reformatted and augmented version of the testimony of James P. Evans M. D., Ph.D, for the Hearing of the House Energy and Commerce Committee's Subcommittee on Oversight and Investigations, dated July 20, 2010.

Chapter 4 is an edited, reformatted and augmented version of a Congressional Research Service publication, Report RL33832, dated January 29, 2010.

INDEX

A

adenine, 54
adverse event, viii, 35, 36, 37
advertisements, 38, 40
African Americans, 13
age, vii, 1, 3, 5, 6, 18, 19, 34, 48
allele, 34, 55
American College of Medical Genetics, viii, 6, 45
amino acids, 54
arrhythmias, 26
arthritis, 18, 21, 23, 25, 27, 29, 31, 33
asthma, 23, 27, 34
athletes, 3, 17, 22
atrial fibrillation, 5, 8, 11, 14, 34
authority, 36, 37, 38, 39, 42

B

background information, 4
behavioral change, 12, 47
behaviors, 12
benefits, 53
biological processes, 18
blood, 16, 34, 47
bone, 21, 27
brain, 3
breast cancer, 3, 5, 8, 20, 26, 53, 57
business model, 39, 41

C

calibration, 48
cancer, viii, ix, 3, 4, 16, 23, 24, 25, 26, 27, 28, 29, 30, 31, 32, 33, 34, 41, 45, 47, 51, 52, 55
cardiovascular disease, 47
CDC, 4, 9, 41, 63
Center for Devices and Radiological Health, viii, 35, 41
chemical, 21, 54, 59, 64
chemotherapy, 59
childhood, 34
children, 3, 17, 19, 34
cholesterol, 18, 21, 23, 34, 47
chromosome, 34, 54, 55
cigarette smoke, 21
CLIA, 37, 38, 42, 61, 62, 63, 64
climate, 53
clinical diagnosis, 56
colon, 4, 5, 11, 14, 15, 16, 23, 34
commercial, 38, 39, 40
confidentiality, 49
consent, 12, 13, 14, 17, 20
consumers, vii, 1, 2, 3, 4, 5, 6, 7, 9, 11, 12, 13, 14, 15, 16, 17, 18, 19, 20, 34, 38, 40, 41, 42, 61

cosmetics, 38
cost, 21, 58, 62
counseling, 16, 34
credentials, 38
cure, 3, 16, 17, 18, 20, 36, 38
cystic fibrosis, 10, 52, 57
cytosine, 54

D

data analysis, 39
dementia, 23
deoxyribonucleic acid, 54
Department of Health and Human Services, 19
diabetes, 4, 12, 25, 26, 27, 29, 31, 33, 34, 52
diet, 15, 18, 19, 23, 34
direct-to-consumer (DTC) genetic tests, vii, viii, 1, 3, 35
disability, 53
disclosure, 20
discrimination, viii, 51, 56, 58, 59
diseases, vii, ix, 1, 3, 5, 7, 10, 11, 13, 15, 17, 19, 21, 23, 36, 39, 41, 47, 51, 52, 53, 54, 55, 56, 57, 64
DNA, vii, 1, 2, 3, 5, 7, 9, 11, 13, 14, 17, 18, 19, 20, 21, 22, 34, 49, 54, 55, 56, 57, 59, 60, 61
doctors, 12, 15, 18, 43, 46
donors, vii, 1, 2, 4, 5, 7, 8, 10, 11, 12, 14, 15, 16, 21, 22
dosing, 53, 58, 59
double helix, 54
drugs, 26, 38, 46, 60

E

economic consequences, 61
education, 37, 53
employment, viii, 51, 58, 59
endorsements, 3, 17, 18, 22
enforcement, 39, 40, 48
environment, 52, 60, 64
environmental factors, ix, 51, 52, 53, 55
enzyme, 21, 54, 59

ethnicity, vii, 1, 5, 6, 7, 12, 13, 14, 23
exercise, 15, 16, 18, 19
expertise, 39

F

false negative, 37
famil(ies), 20, 38, 53, 59
family history, 11, 13, 16, 19, 23, 26, 27, 34
fat, 34
Food and Drug Administration (FDA), viii, 3, 4, 18, 21, 22, 35, 36, 37, 38, 39, 40, 41, 42, 43, 48, 62, 63
federal government, 9, 62
Federal Trade Commission Act, 37
fertilization, 57
fibrillation, 11, 24, 26, 28, 32, 34
fibrosis, 57
field of genetics, viii, 6, 45
flexibility, 63
formula, 18, 22

G

GAO, vii, 1, 2, 3, 4, 9, 10, 17, 20, 25, 27, 29, 31, 33, 34, 40, 46
genes, viii, 13, 20, 22, 34, 36, 38, 45, 52, 54, 55, 56, 57, 58, 59, 64
genetic factors, 38
genetic information, viii, 6, 16, 45, 47, 49, 53, 57, 58, 59, 64
genetic marker, 10, 16, 59, 64
genetic predisposition, 42
genetic testing, ix, 4, 5, 6, 10, 17, 20, 38, 40, 46, 51, 52, 53, 54, 56, 57, 58, 59, 60, 64
genetic tests, vii, viii, 1, 3, 4, 7, 12, 19, 35, 38, 39, 40, 41, 42, 43, 46, 51, 53, 56, 57, 58, 59, 60, 61, 62, 63
genetics, vii, viii, ix, 2, 3, 6, 10, 36, 45, 46, 51, 52, 53, 54, 56, 64
Genetics in Medicine, viii, 6, 45, 64
genome, viii, 8, 38, 45, 46, 52, 55, 64
genomic technology, viii, 51
genotype, 18, 55, 58
guidelines, 18, 46, 49

H

health, vii, viii, ix, 1, 4, 7, 11, 12, 15, 17, 18, 20, 21, 22, 26, 27, 35, 36, 37, 38, 39, 41, 42, 46, 47, 48, 49, 51, 52, 53, 57, 58, 59, 60, 61, 62
health advice, vii, 1
heart attack, 5, 8, 11, 14, 15
heart disease, ix, 4, 11, 26, 27, 38, 47, 51, 52, 55
heart rate, 26, 34
height, 55
hepatitis, 37
heritability, 64
HHS, 62
high blood pressure, 18
history, 11, 23, 26, 27, 34, 43
hormones, 54
human development, 54
human genome, viii, 45, 54
hypertension, 2, 4, 5, 8, 23, 34

I

immune disorders, 4, 55
immune system, 34
in vitro, 36, 57
insulin resistance, 34
integration, 62
internist, viii, 45
intervention, 41, 57

L

labeling, 40
laboratory tests, viii, 37, 51
law, 20, 38
legislation, viii, 51
leukemia, 5, 8, 54
lifestyle changes, 12, 16
liver, 34, 37
low risk, 12
lung cancer, 27
luteinizing hormone, 37

M

management, 38, 53, 61
manufacturing, 42
market segment, 40
marketing, vii, 3, 6, 17, 40, 41, 42, 58, 61
Medicaid, 37, 61
medical, viii, 2, 3, 5, 6, 7, 11, 16, 34, 35, 36, 38, 41, 42, 45, 46, 48, 49, 53, 59, 62, 64
Medicare, 37, 61, 62, 64
medication, 18, 41, 42, 57, 58
medicine, viii, ix, 36, 43, 46, 49, 51, 52, 53, 60, 62
Mediterranean, 15
melanoma, 4
mental illness, 55
metabolites, 56
metabolizing, 59
methylation, 3
molecular biology, 56
multiple sclerosis, 5, 8
mutation, 38, 48, 55, 56, 57, 59

N

National Institutes of Health, 18, 21, 22, 35
National Research Council, 64
nucleic acid, 56
nutrients, 18, 22
nutrition, 3

O

obesity, 4, 5, 12, 14
ovarian cancer, 48
ovaries, 48
overweight, 12, 23, 27

P

pancreas, 34
pathologist, 39
patient care, 40, 46
peer review, 15

pharmacogenetics, 42
phenotype(s), 55, 56
physical characteristics, 55, 64
physicians, 5, 7, 16, 17, 19, 39, 61
policy, viii, 51, 53, 56
predisposition to cancer, viii, 45
prevention, 15, 16, 38
promote innovation, 40
prophylactic, 53
prostate cancer, 2, 5, 8, 9, 34, 47
proteins, 56
public health, viii, ix, 51, 52
Public Health Service Act, 37
public policy issues, viii, 51, 52,53, 56, 60, 61

Q

quality assurance, 63
quality control, 38, 63
query, viii, 45
questionnaire, 19

R

race, vii, 1, 5, 6, 12, 13, 14
radiation, 59
reagents, 36, 42, 62, 63
reality, viii, 4, 45, 46, 47
recreational, 26
recruiting, 21
regulations, 38, 47, 48
regulatory framework, 42
reliability, viii, 2, 4, 6, 17, 19, 36, 49
repair, 3, 18
replication, 59
researchers, 8, 52
response, 20, 34, 42, 57, 58
rheumatoid arthritis, 5
risk, vii, viii, 1, 2, 3, 5, 7, 8, 9, 10, 11, 12, 13, 14, 15, 16, 17, 20, 21, 23, 28, 29, 30, 32, 34, 35, 36, 37, 38, 39, 40, 41, 42, 43, 45, 47, 48, 49, 53, 55, 57, 59, 60, 63
RNA, 56

S

saliva, 3, 5, 41
science, 8, 10, 35, 49
scientific publications, 5
sclerosis, 25, 27, 29, 31, 33
serum, 18
sex chromosome, 54
skin, 4, 18
smoking, 15
society, ix, 51, 61
sodium, 37
stakeholders, 42, 43
standardization, 9
structural protein, 21
susceptibility, 24, 25, 26, 27, 28, 29, 30, 31, 32, 33, 57
symptoms, 15, 16, 55, 57, 59
syndrome, 5, 8, 14, 25, 27, 29, 31, 33

T

technology, viii, 45, 46, 47, 51, 53, 56
testing, viii, 6, 10, 13, 14, 15, 17, 20, 34, 37, 38, 40, 46, 48, 51, 53, 54, 57, 58, 59, 60, 61, 63
therapy, 36
thymine, 54
thyroid, 4
tissue, 59
tobacco, 34
training, 4, 37, 63
traits, 14, 23, 34, 41, 42, 55
transition period, ix, 51, 52
translation, 52
treatment, viii, 19, 36, 37, 38, 39, 40, 46, 51, 53, 56, 58, 60, 61
trial, 4
tumor, 59
type 1 diabetes, 5, 8, 11, 26
type 2 diabetes, 5, 13, 14, 23, 34

U

undercover calls, vii, 1, 17
unproven disease predictions, vii, 1
UV, 59

V

venture capital, 49
vitamins, 18

W

well-being, 22
Western Europe, 14
wood, 21

Y

yield, 2, 7
young adults, 34